LED Packaging Technologies

Design, Manufacture, and Applications

Luruthudass Annaniah, Mohamed Salleh M. Saheed, and Rajan Jose

WILEY-VCH

Authors

Dr. Luruthudass Annaniah
OSRAM Opto Semiconductors
Free Industrial Zone Phase 1
Bayan Lepas
11900 Penang
Malaysia

Dr. Mohamed Salleh M. Saheed
Infineon Technologies
Kulim Hi-Tech Park
Industrial Zone Phase II
09000 Kulim, Kedah
Malaysia

Prof. Rajan Jose
Universiti Malaysia Pahang
Faculty of Industrial Sciences & Technol
Lebuhraya Tun Razak
26300 Kuantan, Pahang
Malaysia

Cover Images: Night view of Kek Lok Si Temple, Penang Malaysia, Photo by Dr. Luruthudass Annaniah
(inset images) Courtesy of ams OSRAM Group

Library of Congress Card No.: applied for

British Library Cataloguing-in-Publication Data
A catalogue record for this book is available from the British Library.

Bibliographic information published by the Deutsche Nationalbibliothek
The Deutsche Nationalbibliothek lists this publication in the Deutsche Nationalbibliografie; detailed bibliographic data are available on the Internet at <http://dnb.d-nb.de>.

Print ISBN: 978-3-527-34878-7
ePDF ISBN: 978-3-527-83166-1
ePub ISBN: 978-3-527-83168-5
oBook ISBN: 978-3-527-83167-8

Cover Design: SCHULZ Grafik-Design

Typesetting: Straive, Chennai, India

Contents

About the Authors

Luruthudass Annaniah is the Director of Product Development at ams OSRAM Penang. He was also an Adjunct Lecturer at University Technology Petronas and was occasionally invited for technical talks and as an external examiner for the final year project of the School of Physics. He has 30 years' of experience in LED packaging technology. He started his career at Siemens Opto Semiconductors as a Product Development Engineer in the year 1993 and grew together with LED evolution at Siemens and then at OSRAM. He has held many positions at Siemens and OSRAM. First as Product Development Engineer, then as Product Development Manager, Senior Engineering Manager, Senior Product Development Manager, and finally as Director of Product Development. In his younger days, he developed Radial LED, then the Surface Mount LEDs, Photodiode LEDs, Laser Diodes and Sensors for Automotive, Consumer, and Industrial product applications. He also holds several patents in LED packaging and has published many technical papers related to LED packaging technology. He holds a Bachelor of Engineering in Mechanical and Material Engineering from the National University of Malaysia, Bangi, Selangor. A Masters degree in Business Administration from University of Strathclyde, Glasgow, United Kingdom, and a PhD in Applied Physics from University of Sains, Malaysia.

Mohamed Salleh M. Saheed is currently working at Infineon Technologies Kulim as Technical Project Leader in Technology, Development, and Innovation Department and specializes in Semiconductor Power Devices for automotive and industrial applications. In his younger days, he worked as Process Development Engineer in wire bond and chip bonding at Osram Opto Semiconductor (now known as ams OSRAM) for four years and as Project Manager in later years. In further career development, he pursued his PhD at the Department of Fundamental and Applied Sciences at University of Technology, PETRONAS, where he specialized in nanostructures of electron transporting materials and graphene research and its application in perovskite solar cells (PSC). He also published many technical papers related to semiconductors in P–N junction solar cells and LEDs.

Rajan Jose is a senior Professor at the Universiti Malaysia Pahang (UMP) and the Associate Editor-in-Chief of the Springer Nature journal Materials Circular Economy. He has served as the Dean of Research (Technology) at UMP during February 2016–August 2019, besides serving as the Member of Senate and Graduate Council of UMP. He has investigated nanostructured perovskite ceramics for microwave and superconducting electronics during doctoral research at the Council of Scientific and Industrial Research (CSIR), Trivandrum, India, and received his PhD degree in the year 2002. He has contributed to the science and engineering of diverse range of materials, including inorganic and organic semiconductors, polymers, metals and alloys, materials for molecular electronics, luminescent quantum dots, biomaterials, glasses, and glass ceramics. He was employed as a scientist at the Indira Gandhi Centre for Atomic Research (India), AIST (Japan), Toyota Technological Institute (Japan), and the National University of Singapore (Singapore) before joining UMP. He has published nearly 300 papers in the Web of Science (Thomson Reuters/Clarivate Analytics)-indexed journals, which have been cited nearly 17000 times with an h-index of 66 according to Google Scholar database. He holds 25 patents. He has supervised 6 Postdoctoral, 24 Doctoral, and 10 Master's researchers. Stanford University places him as a top 2% Materials Scientists in the world ever since the list was produced in 2020. His current research interests include sustainable materials, textile electronics, circular economy, data science, and renewable energy devices; most of his research is on the structure – property relationship in materials for a desired device functionality.

Preface

Light-emitting diodes (LEDs) have played a substantial role in our daily lives ever since their commercial use for over 60 years and will continue to grow in significance as they are the present-day solution for energy-efficient and low-carbon lighting. Haitz's Law predicts 20 times increase in LED efficiency per decade, thereby providing strong hints for further cost reduction over time besides new application domains for LEDs. Together with enhancing materials efficiency and circularity, progress in LED packaging technologies in designing and manufacturing processes claims a major role in their commercial success. By using appropriate packaging technologies, the power density handling capacity of LEDs has remarkably increased to 16 W in 2022, up from less than 0.15 W in early 1990s. The revolution in LED packaging technologies enabled their high-power applications and has paved the way for sensing using infrared LEDs (IRLEDs). More new discoveries have been made by employing IRLEDs in applications such as bio-sensing, security, automotive sensing, and many others. Consequently, the LEDs are expected to dominate almost all consumer products and industrial applications thanks to their high brightness and low cost. Environmental robustness of LED packaging has become a critical part of the design and the manufacturing to prolong its lifetime while deployed in harsh environments. It is imperative to understand the fundamentals of LED packaging technologies to bring the best out of LEDs for many newer applications. This book focuses on LED packaging technologies and applications, which are not readily available in the literature, in order to help engineers and scientists develop efficient LEDs and ensure their rapid adoption around the world. This book is organized into nine chapters, as outlined below.

Chapter 1: A Brief History of Artificial Light and LED Packaging. This chapter explains the history of primitive lighting in human civilization to the invention of modern LED-based lighting as well as the evolution of LED packaging technology over the years, including package-free LEDs. In addition, a brief summary of present LED application technologies and anticipated ones is provided.

Chapter 2: Fundamentals of LED Packaging Technology. This chapter explains the basic light emission and extraction technology in LED packages and covers how package design impacts light emission and extraction. The thermal impact on light efficiency is discussed generally. The package thermal design is detailed. Electric contact design is vital for LED packages to avoid series resistance impacting the LED's performance. Here the electrical contact design is briefly explained. Optical design, which is critical for the LED application and affects the LED performance and efficacy, is also briefly outlined.

Chapter 3: LED Package Manufacturing Technology. This chapter explains the LED package manufacturing technology. This covers backend end of LED manufacturing, which explains the die attach, wire bonding, encapsulation, and the LED testing technology. The LED testing technology covers the concept of LED electro-optical testing. The challenges in LED testing technology, especially approaching zero defect expectations at the end customer are also explained.

Chapter 4: LED Automotive Lighting Application Technology. This chapter explains the existing and future LED applications for automotive industry. The lighting system for automotive vehicles plays a critical part in safety of both the driver and other road users **to see** and **to be seen**. The automotive lighting is divided into two categories: **Signaling** lights, which are the lights that are **to be seen**, and the **Lighting** lights, like the headlamps, which are lights **to see**. These automotive lighting applications are explained in detail.

Chapter 5: LED Application for Consumer Industry. This chapter explains some of the existing and future medical industry applications of LED, especially in healthcare treatments. The LED usage in skin rejuvenation, wound healing, and closure was discussed. Bio-sensing to check heart rate and oxygen saturation level in blood to mention a few. Future trends in LED applications pertaining to consumer industry are also explained here, for example, iris recognition for security, spectroscopic sensing for food, and food safety treatment.

Chapter 6: LED Application for General Lighting. In this chapter, the LED application for local illumination for interior and exterior is explained. Horticulture LED application is getting increasingly important and is given much emphasis in this chapter. The concepts and technological deployment of general lighting are explained in detail in this chapter.

Chapter 7: Quantum LEDs. This chapter briefly compares Quantum Dot LEDs with Organic LEDs. A detailed introduction to the quantum confinement phenomena and size-dependent optical properties of quantum dots is presented, and the advancement of future QD LEDs is discussed. The challenges and possible solutions are outlined as well.

Chapter 8: Ultraviolet LED Packaging and Application. This chapter is devoted to UV LED application and packaging technology. It covers the categorization of UV LEDs based on their wavelength and their specific applications. The packaging technologies and challenges based on the UV LED categorization were also explained in detail.

Chapter 9: Life Cycle Analysis and Circular Economy of LEDs. This chapter is devoted to the life-cycle analysis of LEDs, existing recycling protocols, their relative advantages and disadvantages, and circular economy (i.e. sourcing the next-generation LED materials from the existing LED wastes). The role of artificial intelligence, big data, and blockchain technology in future recycling of LEDs is elaborated.

We hope that this book will be a valuable source of reference for all those who are keen to understand LEDs and their ever-expanding applications. We also sincerely hope it will be an aid in further research into this LED packaging technology. The subject covered in this book will give a basic understanding of LED packaging science and its applications. This in turn may inculcate further research and development. It is our hope that the information presented in this book may assist in removing some of the barriers to development of LED packaging and new applications.

Dr. Luruthudass Annaniah
OSRAM Opto Semiconductors
Bayan Lepas, Penang, Malaysia

Dr. Mohamed Salleh M. Saheed
Infineon Technologies
Kulim, Kedah, Malaysia

Prof. Rajan Jose
Universiti Malaysia Pahang
Kuantan, Pahang, Malaysia

Acknowledgments

This book is written based on knowledge gained from working experience of almost 30 years in LED industry and research on LED packaging technology during my tenure in Siemens Opto Semiconductors, Deutsche Technoplast GmbH, and ams-OSRAM.

I am so grateful to ams-OSRAM for giving me the access to so much information and permission to write this book.

I am also grateful to a number of mentors, friends, and colleagues in this work:

In Germany, I am grateful to my mentors, Frank Moellmer, Guenter Waitl, Dr. Raimund Schwarz, Herbert Brunner, and late Dr. Thomas Hoefer for their generous guidance and help in my career. I am also very grateful to my friends and colleagues, Lex Wolfgang, Dr. Jeorg Strauss, Dr. Matthias Sabathil, Dr. Markus Arzberger, Dr. Hans Christoph Gallmeier, Markus Horn, Georg Bogner, Dr. Bernard Stapp, Dr. Martin Strassburg, Dr. Michael Schwind, Dr. Martin Behringer, Reichel Marion, Hubert Maiwald, Thomas Kippes, Hubert Hoelzl, Thomas Schreiber, Christoph Walter, Sigrid Putz, Guenter Heidel, Robert Lutz, Bodo Ischebeck, Thomas Zahner, and many more, for their generous support and friendship.

In Penang, I am grateful to my mentors, Yip Heng Keong, Bala Vythilingam, Khor T.K., and late Datuk Yap for their guidance and help in my career. I am also grateful to my friends and colleagues, Won Yun Sung, Kok Foong Chau, Ludwig Hofbauer, Prof. Mutharasu, Cheng Kai Chong, Chew Chee Wah, Jerry Tan, Dr. David Lacey, Thayalan, Brandon Tan, Tan Wei Jia, Cheah Mun Wai, Jade Looi, Dave Lai Chan Wei, Bernard Raj, Dr. Shanmugan, Vinod Panicker, Samivel, Dr. Lim Chon Kean, Dr. Lim Weng Hong, Purusothaman, Melvin Ho, Teh Beng Hui, Tony Chen, Tan Lean Nee, Chim Weng Yau, and many more for their generous support and friendship.

Finally, I would like to acknowledge with gratitude, the support and love of my family – my wife Maria, daughter Ivy, and son Juan for their invaluable support and love.

They all kept me going, and this book would not have been possible without them.

Dr. Luruthudass Annaniah

In the name of Allah, the Most Gracious, the Most Merciful,

To Dr. Luruthudass, for inviting me and giving me endless research guidance and support to co-write this book,

To my most beloved mother, Marliya Ban,

To my wife Nafsiah Begum and my boys Muhammad Yusuf and Muhammad Idris, my ever-loving family,

To also acknowledge Infineon Technologies for giving me permission to work on this book.

Dr. Mohamed Salleh M. Saheed

I am grateful to my colleagues Dr. Dass and Dr. Salleh for inviting me to work on this project; their rich and long experience with solid-state lighting technologies gave me an opportunity to learn the subject a bit more deeply. I also acknowledge Prof. Seeram Ramakrishna, National University of Singapore, for helpful discussions on material sustainability and materials circular economy.

Prof. Rajan Jose

1

A Brief History of Artificial Light and LED Packaging

1.1 Evolution in Artificial Light

Light is one of the most important ingredients for the survival of all living things. The primates as far back as two million years ago might have used fire from burning wood as artificial light [1]. This highly intelligent primates' survival instinct mastered the usage of burning wood for many other uses than to see and to be seen. The primates in the early years have also learned to make artificial light by making fire by rapidly grinding two combustible materials.

Light has fascinated human beings since the dawn of civilization, and artificial light has played an important role in human civilization. In the 1980s, archeologists unearthed an oil lamp made of stone in a cave in Southern France. The occupant of this cave was using the lamp for cave drawing [2]. This may be the first known lighting tool that uses fat-burning fuel. Carbon dating indicated the lamp might have existed some 38,000 years ago. These lamps were made from limestone or sandstone and can be easily fashioned with shallow depressions to retain the melted fuel. Chemical analysis of residues of the fuel has shown that it was probably animal fat [3]. This Paleolithic lamp, as illustrated in Figure 1.1, has the lighting power of a candle. Oil lamps are still in use today in some parts of the world, where electricity is not readily available or affordable [4]. Civilization has accelerated ever since the invention of artificial light, as their productive hours have extended beyond daylight into the night and even indoor activities [5]. The artificial light based on fuel-burning technology has since evolved from oil to kerosene and gas-discharged lamps [4].

In the nineteenth century, there was a breakthrough in artificial light when electric light was invented. Electric light or the incandescent light bulb was further perfected by Thomas Alva Edison [6]. However, this light source was very inefficient as it converted less than 5% of the energy to light and the rest was turned into thermal energy. In the early twentieth century, fluorescent and sodium lights took over the standard incandescent light bulb. However, this light source has its issues such as the content of hazardous materials like mercury and short product life span [4]. Hence, this allows the light-emitting diode (LED) to shine as it offers an alternative way of light generation. LED's spontaneous light emission due to radiative recombination

LED Packaging Technologies: Design, Manufacture, and Applications, First Edition.
Luruthudass Annaniah, Mohamed Salleh M. Saheed, and Rajan Jose.

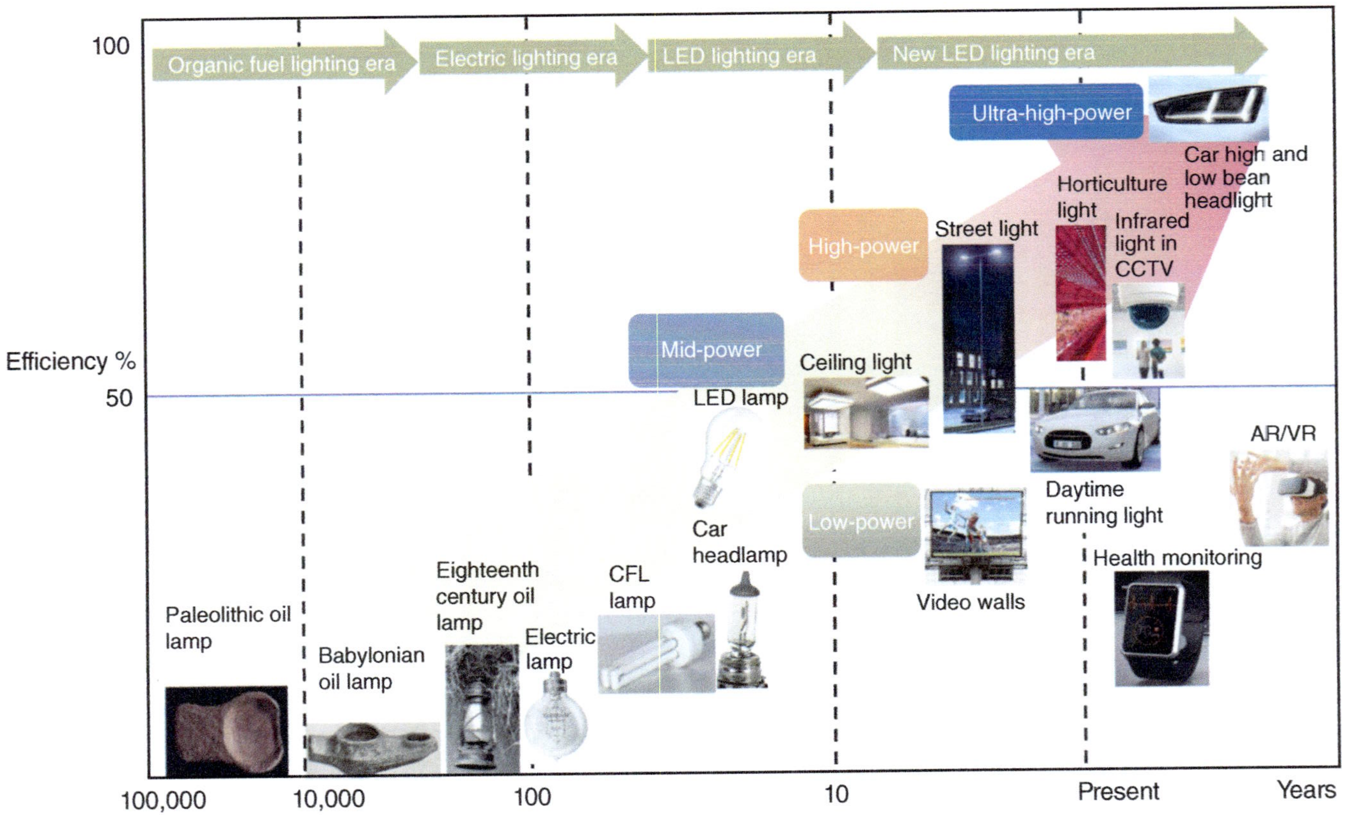

Figure 1.1 Evolution of artificial light.

of excess electrons and holes is an important selling point that attracts a lot of interest besides their energy efficiency.

Even though LED was discovered earlier than compact fluorescent light, it did not flourish as there was not much development or innovation in the early years. LED was first discovered by Henry Joseph Round in 1907. He found Silicon Carbide (SiC) illuminates when it is biased with 10 to 110 V. This early form of LED was very dim. In 1928, Oleg Vladimirovich Losev, a brilliant inventor and genius physicist, reported a detailed investigation of the luminescence phenomenon observed with SiC metal–semiconductor rectifiers. He found the light could be switched "on" and "off" rapidly, making it suitable for what he called "light relays." His discovery of crystaldyne, which was the first crystal amplifier and oscillator, and the invention of the first semiconductor LED generating visible light could be the basis for the development of semiconductor electronics. However, this SiC had an efficiency of only 0.03% and was not comparable to the current III–IV material system. In the late 1950s, Welker's [7] proposal suggested that compound semiconductors from III and V groups of the periodic table should have comparable semiconductor properties to those of germanium (Ge) and silicon (Si). These led to the discovery of infrared (IR) emission from gallium arsenide (GaAs) crystals with very low quantum efficiencies of around 0.01–0.1%. This early observation and understanding of band structures of semiconductor materials were soon followed by the quest for visible LED. This is where Nick Holonyak and Bevacqua invented the red LED in 1962 [8, 9]. They were using vapor-phase epitaxy (VPE) of gallium arsenide phosphate (GaAsP) on a GaAs substrate. This technique was used to produce the first red luminescence diode, triggering an industrial production revolution in LED manufacturing, where many applications like indicator lights and alphanumeric displays benefited [7]. Monsanto Corporation was the first to start commercial mass production of LED in 1968. It produced low-cost GaAsP LEDs. Hewlett–Packard (HP) Corporation joined the race to develop LEDs in the late 1960s, followed by other corporations [10]. Development of new semiconductor materials has made it possible to produce LEDs in a variety of colors as they become even more effective for use. However, high-brightness and efficient blue LEDs based on gallium nitrate (GaN) came in the early 1990s. Isamu Akasaki, Hiroshi Amano, and Shuji Nakamura made it possible to obtain very efficient blue and green LEDs [11]. This led them to win the Nobel Prize for Physics in 2014. The invention of efficient blue LEDs has enabled white-light illumination. In 1997, white light was demonstrated for the first time by combining a blue GaN LED with a yellow-emitting phosphor [12], which revolutionized solid-state lighting (SSL).

Figure 1.2 shows commercialized and widely used SSLs such as traffic signals, backlighting for screens, televisions, video walls, interior and exterior lighting for automobiles, home lighting, stage lighting, mobile phones, and many more applications. With continuous improvement in the performance and cost reduction in the last decades, SSL is replacing conventional light rapidly.

In comparison to conventional lighting, SSL has two highly desirable features: (i) energy-efficient and consequent reduction of carbon footprint and (ii) extremely versatile with many controllable properties, including the emission spectrum,

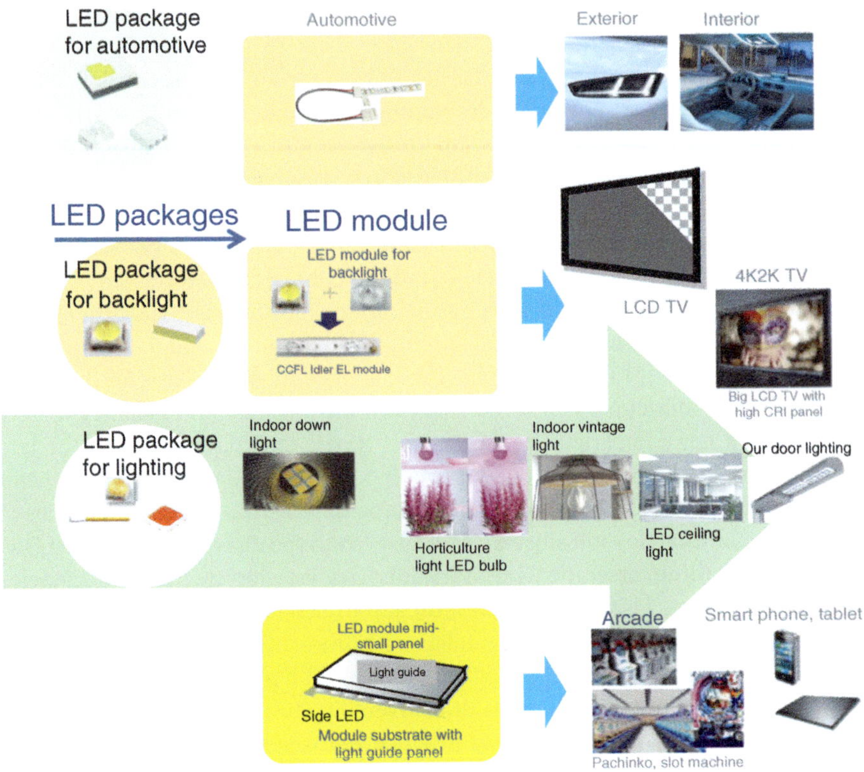

Figure 1.2 Solid-state lighting expanding application. Source: Courtesy of ams OSRAM GmbH.

direction, color temperature, modulation, and polarization. The impact of LEDs on the economy, environment, and quality of life has become very significant.

1.2 Impact of Light-Emitting Diode on the World

The invention of the GaN-based blue LED has significantly transformed the lighting industry. In the last decade, LED light sources have gone from being just an interesting novelty to a new light source option that can be used for energy savings, longer lifespan, and higher performance in almost any application. For example, a 15-W LED lamp can replace a 75-W incandescent lamp, deliver a useful lifetime averaging 25,000 hours, have adjustable lighting, require no warm-up time, and offer superb color rendering [13, 14].

LED provides significant energy savings because it converts energy efficiently compared to other light sources. The cost-saving advantages are revealed in a study by Ehrentraut and Meissner in 2010 on the impact of the conversion to SSL on US electrical energy consumption, as illustrated in Table 1.1. In this study, the energy consumption estimated by assuming the power consumption of solid-state light to produce almost identical light output is just a fraction compared to a 60-W

Table 1.1 Potential impact of conversion to solid-state lighting on U.S. electrical energy consumption.

General illumination lighting			
Performance estimates	**SSL product**	**60-W incandescent light bulb**	**23-W compact fluorescent lamp**
Light output (lm)	1000	1000	1200
Power (W)	6.67	60	23
Lumens/W (system)	150	16.7	52
Annual energy consumption (8 h/d, 365 d) (kWh)	19.5	175.2	67.2
Factor higher than LED	1	9	3.4
Annual energy cost per lamp (9.3 ¢/kWh)	US$ 1.81	US$ 16.29	US$ 6.25
Estimated annual energy savings with LED lighting: 2020 estimated baseline energy consumption for lighting: 7.5 quads			
% US lighting conversion to SSL	**Quads saved**	**$ saved (billions)**	
1%	0.05	0.33	Assumption: equal replacement of incandescent and fluorescent lighting
10%	0.49	3.23	
25%	1.21	7.99	
50%	2.43	16.04	

"quad" is one quadrillion BTU; approximately US$ 6.6 billion per quad of electrical energy.
Source: Ehrentraut et al. [13]/Springer Nature.

incandescent light bulb or compact fluorescent lamp (CFL). The efficiency of SSL products is almost three times better than CFL products and nine times better than that of an incandescent light bulb. Annual energy cost per lamp can be estimated as energy cost is 9.3 ¢/kWh, SSL product energy cost is US$ 1.81 compared to an incandescent light lamp of US$ 16.29 and a CFL product at US$ 6.25. These advantages alone captured much attention and ensured a strong future for LEDs [15]. As a result, this led to the LED industry's double-digit growth over the last decades.

The global LED lighting market size was valued at US$ 54.00 billion in 2019 and is projected to expand at a compound annual growth rate (CAGR) of roughly 11%. Growing stringency of regulations in terms of inefficient lighting technologies and rising government efforts toward sustainable development are the key growth drivers. An aggressive decline in the prices of LED, coupled with the transformation in energy policies across the world, has been a driving mechanism for market growth. Moreover, attractive incentives and rebates provided by the governments for the use of LED lighting in several countries will leverage the demand.

LEDs are highly efficient and reliable, and they yield a longer life span, which is anticipated to boost their application in both indoor and outdoor settings. These lights, at present, are the most cost-effective light, compared to incandescent lights, delivering around 100,000 hours of illumination with a small amount of energy consumed [16]. Their lower cost of operation and reduced heat losses make them a suitable alternative for incandescent lights. Technological advancements shift from conventional to green lighting, enhanced energy efficiency standards, and declining prices have also spurred product demand. As a whole, this clearly shows that the LED business had all the potential for explosive expansion. Many new inventions were seen in LED products and application technologies. Revolutionary LED growth in many areas can be seen as governmental and industrial funds were injected into technology research and development, thus whole supply chain of LED industry is expanding. In the Section 1.3, the LED industrial chain is further elaborated.

1.3 LED Industrial Chain

The LED industry can be divided into three sectors as shown in Figure 1.3. They are Front End (FE), Back End (BE), and LED Luminaire (LL) industries.

As illustrated in Figure 1.3, the FE industry consists of epitaxial growth and chip processing. Here, a thin layer of light-emitting material was grown on a suitable substrate, like sapphire for gallium nitride (GaN), gallium arsenide (GaAs) for aluminum indium gallium phosphate (AlInGaP), and aluminum indium gallium arsenide (AlInGaAs). Mostly, chips are grown in metal–organic chemical vapor

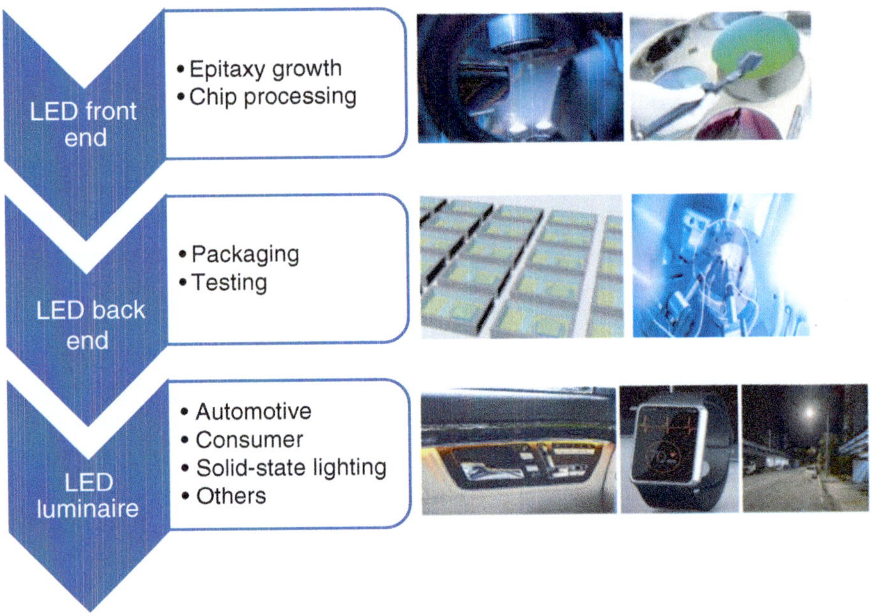

Figure 1.3 LED industrial chain. Source: Courtesy of ams OSRAM GmbH.

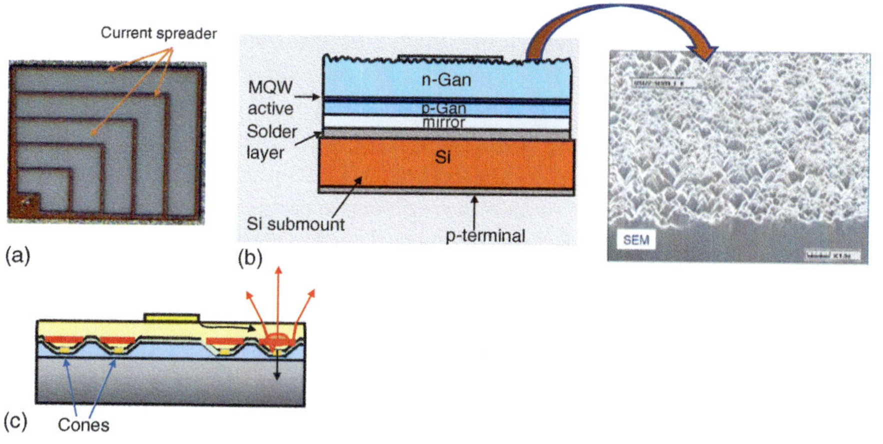

Figure 1.4 (a) AlInGaN chip top view showing current spreader, (b) AlInGaN chip cross-section view and roughened surface of chip surface, (c) AlInGaP and AlGaAs Thinfilm chip cross-section view showing reflector cones. Source: Courtesy of ams OSRAM GmbH.

deposition (MOCVD) reactors. The grown epitaxy wafers are further processed in chip processing processes where current spreaders are sputtered with gold layer to make an electrical connection at epitaxy n-layer that enables to bond for wire-bond at chip, as illustrated in Figure 1.4a.

For the AlInGaN chip, as illustrated in Figure 1.4b, the layer above epitaxy was roughened to improve light outcoupling. A passivation layer is introduced on top of this layer for protection. On the other hand, the AlInGaP and AlGaAs films grown on GaAs substrate go through a different process, especially those thin film technology chips in OSRAM, as illustrated in Figure 1.4c. To improve light extraction, reflector cones were fabricated in complex chip manufacturing processes. Right below these cones is a mirror to improve the reflection. The chips will be 100% tested and inspected for any defects before they are singulated into single chips. Once completed singulation, these wafers will be sent to Back Eng (BE) to package them into an LED product.

LED packaging processes consist of assembling the chip onto a package substrate using conductive or nonconductive glue, which is referred to as chip-bonding. It depends on the chip technology. Thin film chips are commonly glued using conductive inks, while sapphire chips are mostly glued using nonconductive glue. Flip chips, on the other hand, are soldered directly onto the package substrate. A package substrate can be premolded copper frame, ceramic, printed circuit board (PCB), or even metal can. Selections depend on the products end applications and thermal management requirements. Once the chip is attached to the package substrate, the anode is connected to the package by wire bonding process. In LED industries, gold wire was the preferred choice compared to silver or aluminum wire. Flip chip does not need such a wire bonding process, as its anode and cathode are directly soldered to package substrate. Packages with chip and bonded wire require physical protection; therefore, requiring encapsulation process. Encapsulation materials such as silicone, epoxy, or cap made of epoxy or glass will be used. Most of the package substrates are in panel or reel-to-reel (R2R) form. This substrate will be singulated to individual package in the singulation process after the encapsulation process. Here, the packaged LED is either diced via laser cutting or cut mechanically (either by sawing or trimming and forming) to make them into a single LED package for the next process, which are optical and electrical testing. All LED production is subjected to 100% testing and visual inspection to remove defective parts. The final process in BE is packing process. Here, the LEDs are either packaged in reels or trays, and they are vacuum-sealed in moisture barrier bags before being shipped to LED customers. Further details of the BE process will be explained in the LED manufacturing chapter later (Chapter 3).

LED is used in almost every industry, namely, in SSL, automotive, video walls, mobile phones, signage, medical industries, and many other applications. A few selected LED application industries are elaborated in Chapters 4–8.

1.4 Evolution in LED Packaging Technology

Ever since the industrialization of LED in the 1960s, the LED packaging has been evolving in synch with the application technology. It is revolutionized by the application itself. In the beginning, during late 1960s, LED was mostly used in signal lighting and indicators, for example, in signboards and calculators. The LED chips are packaged in lamp or Radial packages, as shown in Figure 1.5. However, the application is limited due to size and heat power dissipation capability of Radial package. Given this weakness, the Surface Mount Device (SMD) LEDs were born in early 1990s.

This opens up more new applications, especially in the automotive and consumer industries. With the invention of the super bright blue LED in the early 1990s and coupled with phosphor conversion technology from existing CFLs, white LED was invented by Nichia and OSRAM, thereby making tremendous impact on the LED lighting industry, especially in the general lighting application. There was an urgent drive to replace conventional lighting with LED-based lighting, which further provoked another wave of drive, in the packaging industry, as the applications demand higher brightness, specific radiation pattern, and variety of sizes. Hence, more variety of package designs were invented to meet the demands, as illustrated in Figure 1.5, where high-power packages and low-power packages evolved. As the LED lighting industry grows, another wave of innovation is triggered to drive miniaturization of the whole system in an application. This drives to integrate the LEDs with other components like ICs, lenses, diodes, and many other small components into one single package that calls System in Package (SiP). In parallel to the SiP, another drive happening while writing this book is to integrate at chip level, called System In Chip, SIC.

The LED packaging offers significant challenges for innovative designs to cater to customized applications. To meet these applications, one has to carefully carve the

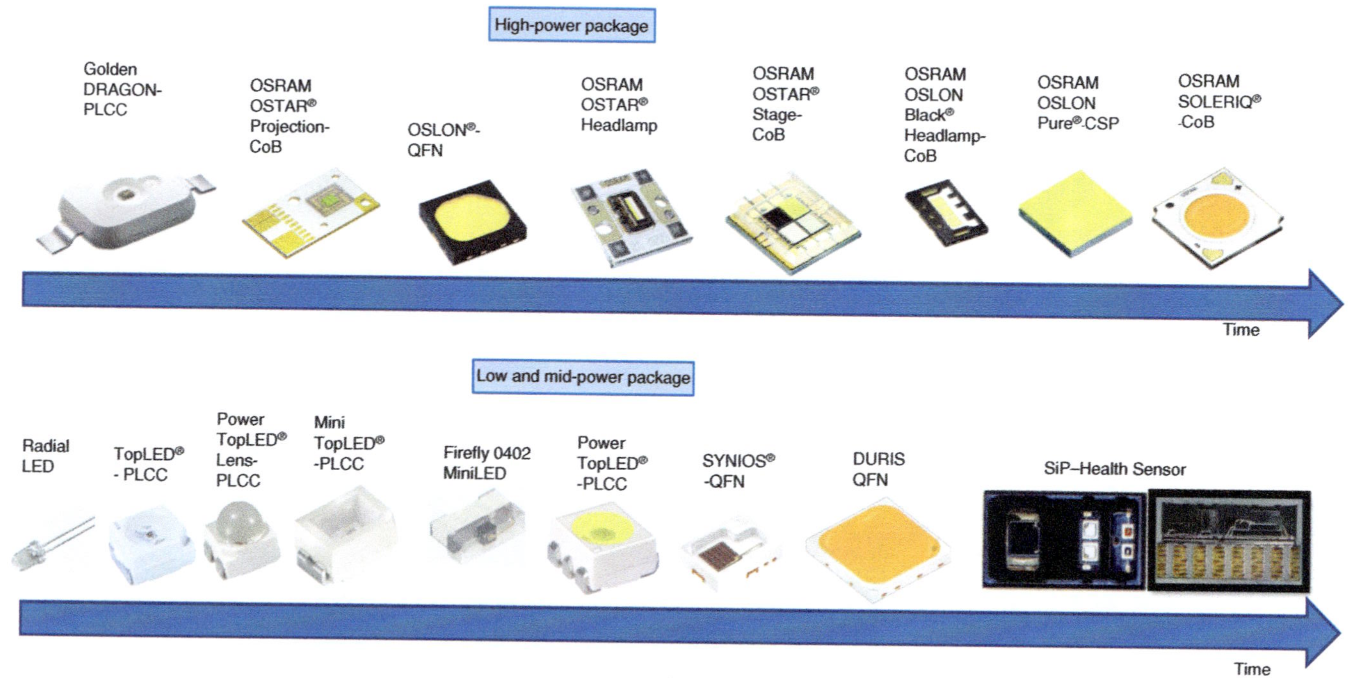

Figure 1.5 LED package evolution. Source: Courtesy of ams OSRAM GmbH.

design by considering the multidimensional space of options between the key components of a package, for example, the package substrate, interconnect, phosphor, optical interfaces, and in some cases, the special features of material like titanium oxide (TiO_2) layer.

The key success of an LED lies in the packaging technology that is able to fit the chip into specific applications. In fact, without a good package, the LED cannot be efficiently used for many applications. Why are the LED packages very important? As illustrated in Figure 1.6a,b there are many functions behind LED packaging. One of them is to interconnect interface with the customer. Second, it is the chip interconnect that is silver or gold-filled glue that adheres chip to the substrate (in this case, premolded package) and gold wire to connect the electrical circuit in the package. These chip, wire, and interconnect glue are sensitive to the environment, hence they need to be protected by the encapsulant. This encapsulant is made of either epoxy or silicone, which protects the chip from the environment and mechanical damages. Thirdly, the LED package offers the optical interface to the customer application.

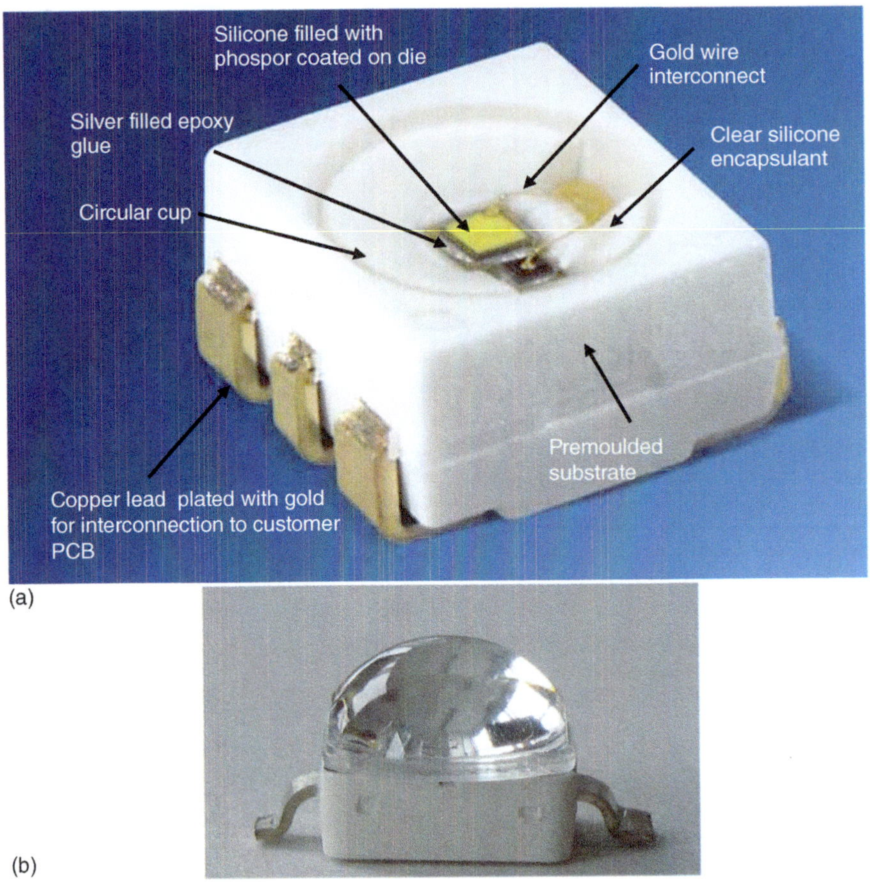

(a)

(b)

Figure 1.6 Illustrate a standard surface mount LED package. Source: Courtesy of ams OSRAM GmbH. (a) Power TopLED® – PLCC, (b) Golden Dragon® Lens – PLCC.

Table 1.2 General description of low-, mid-, high-, and ultra-high-power LEDs packages is in watts.

Low-power	Mid-power	High-power	Ultra-high-power
<0.2 W	0.2 W 1.0 W	1.0 W 4.0 W	>4.0 W

Source: Courtesy of ams OSRAM GmbH.

In Figure 1.6b, the cap's reflective properties shaping the radiation pattern are illustrated. In some packages, the lens on the top gives a unique radiation pattern for a specific application, for example, facial recognition in mobile phone security. It is important to identify the most competitive option for a distinct design-in, i.e. fitting to the technology roadmap of the customer application.

The packaging technology generally can be described in four categories, as in Table 1.2. They are low-power, mid-power, high-power, and ultra-high-power LED packaging. The electrical input power of these LEDs defines these categories, although without any governing standards. In general, LED package is considered as high-power when the electrical power is >1 W. But in a different case, chip sizes greater than 1 mm^2 were categorized as high-power. Mid-power is in the range of 0.2 to <1 W and the low-power <0.2 W.

In the end of 1960s to 1980s, most of the LEDs are in the form of Radial Through Hole LED package or metalcan package. It has its limitations in terms of size and thermal resistance. The LED chip has not matured enough for high drive current as the chip size is small and internal efficiency is still low. The LED was also not so bright. Furthermore, this package requires a through-hole PCB to solder to connect to the customer PCB circuit. This increases the size of the PCB. Hence, the end product or system size also follows the PCB size, which is relatively large. With all these limitations, much new research was taken place at some companies to develop a better package to meet the market demand. This was where the SMD LED was born. The SMD packaging in real sense was already in use in the non-LED semiconductor industry. As a result, there were many cross-learnings from the semiconductor industry to the LED packaging industry.

The SMD package technology has taken the LED to a new level of application. With the SMD package, the size, external lens design, and thermal management design can be more effective to bring in much value to the new application. As an example, having a cup and lens in the package will enable it to have better efficacy compared to no cup or lens, as the light will stray away or absorbed by the package and chip. On the other hand, with better heat sink, the LED package can bring down

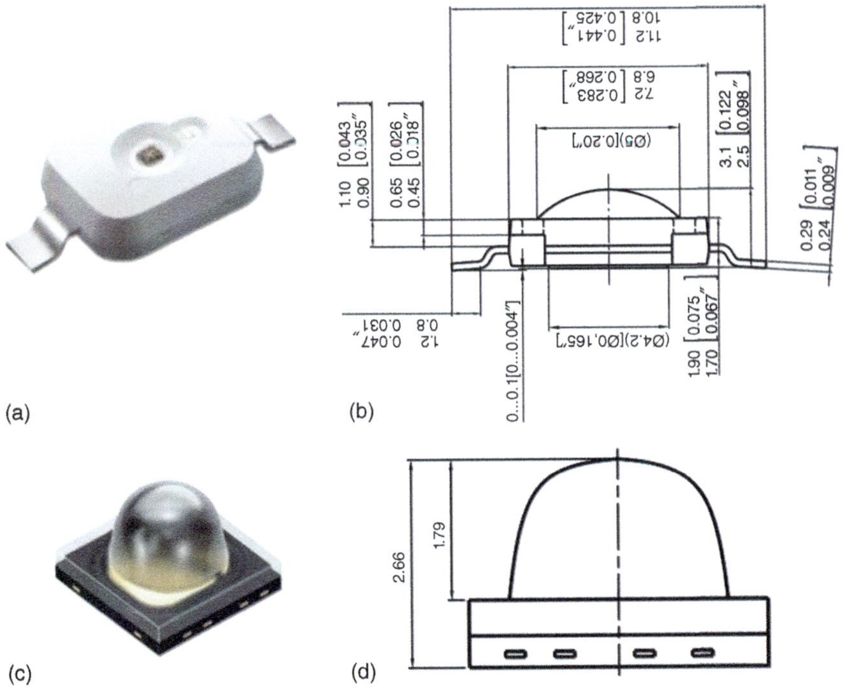

Figure 1.7 High-power packages: (a) Golden DRAGON package, (b) Golden DRAGON package schematic, (c) OSLON Black® package, and (d) OSLON Black package schematic. Source: Courtesy of ams OSRAM GmbH.

the real thermal resistance value significantly low. Example Golden DRAGON® of OSRAM in Plastic Leaded Chip Carrier (PLCC) package and OSLON® Black Series of OSRAM in the Quad Flat No-leads (QFN) package, as illustrated in Figure 1.7. They have real thermal resistance at typical values of 6.5 and 6 K/Wm^2, respectively. This enables the LED to be driven with high current and reap high efficacy and brightness of the LED without affecting the lifetime of the LED.

1.4.1 Low-Power Package Evolution

As mentioned earlier, the term low-power LED package referred to LED with electrical input power less than 0.2 W. These LEDs are usually driven at 5–50 mA with bias voltage in the range of 1.5–3.5 V. This voltage depends on its chip technology. Low-wavelength chips (GaN) have higher voltage compared to higher wavelength (AlInGaP and AlInGaAs) chips. For example, infra-red (IR) chips have voltage in the range of ~1.5 V and blue ~3.5 V. The chip size is usually in the range of 180–300 μm. The usual package constitutes of low-power LEDs are Radial lamp, small SMD top or side view PLCC packages, and ChipLED. The major applications of this package are mobile phone keypad and LCD backlighting. Besides, they are also used for backlighting for computer screens and TVs, signage, and large displays.

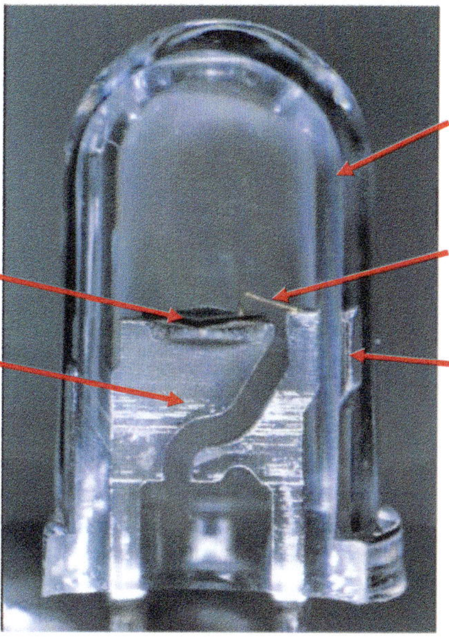

Figure 1.8 Radial package in a cross-section view.

In early days, most of the low-power LED packages are Radial packages as illustrated in Figure 1.8, which shows a simple LED package where the chip was bonded on copper lead and a gold wire connected between the chip and an anode lead. They are encapsulated by an epoxy material. The heat dissipation is mostly through the cathode lead. This is inefficient, however, sufficient for low-power applications.

The Radial LED package design could not progress well in the LED packaging technology due to its limitations. A smaller version of Radial LED was invented by Siemens Opto Semiconductors with narrow radiation pattern specifically for niche applications in sensor applications, for example, smoke detector sensors. In the early 1990s, Siemens Opto Semiconductors developed an SMD LED package, which is friendlier for the surface mount soldering process. It is a compacted LED package with a size of 3.4 mm by 3 mm and a height of 2.1 mm, and it improved the light out-coupling by a factor of 2 compared to Radial LED. It is cheaper than the Radial package. This was a breakthrough in an LED application. As brightness improved, lower cost and thermal resistance relatively lower than Radial package were achieved, and many newer applications of LED came. They are in automotive, traffic light signals, compact signboard, and handheld devices. Low-power LED packages in automotive are mostly used for brake lighting, signal lighting, and dashboard lighting. In the handheld device market, for example, handphones, in the mid-1990s, companies like Nokia, Siemens, Samsung, and Motorola drove the low-power LED package to a smaller size package. The low-power LED package size like Mini TopLED shrank to 2.3 mm by 1.9 mm with a height of 1.4 mm and ChipLED to mere 1 mm by 0.3 mm with a height of 0.57 mm for mobile phone keypad applications in the mid-1990s, before the smartphone era. This size further shrank as the application in Video Wall requires high pixel quality, where LED size and pitch size play an important role in Video Wall color gamut quality.

1.4.2 Mid-Power LED Packages

Mid-power LED packages are those packages that drive current from 50 to 150 mA and bias voltage between 1.5 and 3.5 V. The LED power is generally less than 1 W. Chip size is usually in the range of 300–600 μm. The package design is mostly PLCC packages like SMD top and side view LEDs and ChipLED. Some of the mid-power LEDs are shown in Figure 1.9. It comes in many sizes and forms that fit into the product application at the customer end.

Most of the applications are on TV, backlighting monitors, and automotive lighting on dashboard and interior lighting use mid-power LEDs. It is worth mentioning that many general lighting products use mid-power LEDs. This package can deliver high efficacy with manageable thermal loading. Their lower power consumption often allows standard PCB material, such as FR-4 PCB, to be used compared to the high-power LED package that requires metal-core PCBs. Figure 1.10 shows the mid-power thermal management design evolution. The thermal resistance or conductivity can be reduced significantly by having the heat sink at the package.

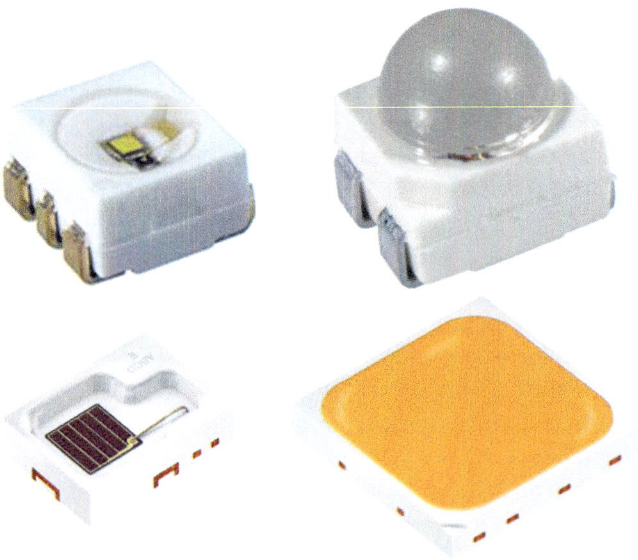

Figure 1.9 Some of the mid-power LED packages. Source: Courtesy of ams OSRAM GmbH.

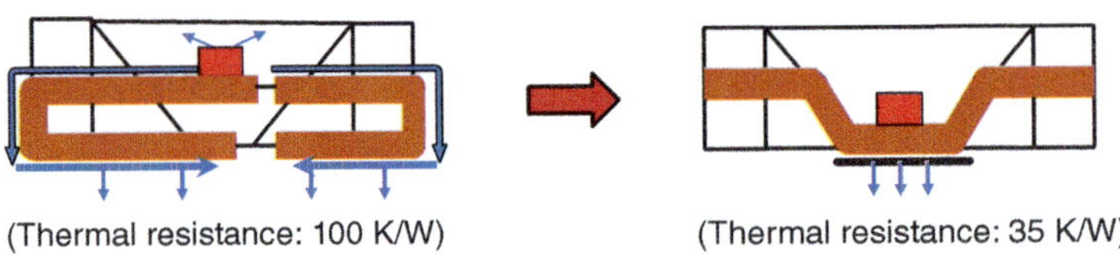

Figure 1.10 Mid-power LED PLCC package thermal management design evolution.

Many LED manufacturers, such as OSRAM Opto Semiconductors, Lumileds, Everlite, and Nichia, are significantly betting on this package category, and the cost of manufacturing in this package is significantly lower than high power, and indeed the market is significantly large.

1.4.3 LED High-Power and Ultra-High-Power Packages

The high-power LED packages are those with input power in the range of 1–4 W, and ultra-high-power packages are those with input power greater than 4 W. Their driving current is greater than 350 mA, and biasing voltage is in the range of 1.5–3.5 V. Chip size is usually in the range of 600–2000 μm. They can have either single large like OSRAM OSONIQ® P 7070 chip or multiple large chip like OSRAM Duris® S8. Some packages have multi-small chips arranged in an array in one package, for example, OSRAM SOLERIQ® S9. There are many types of packages namely single large chip packages; some are multi-chip packages. The package substrate varies. Some are premolded leadframe, ceramic, and metal-core PCB. Figure 1.11 shows the different varieties of high-power and ultra-high-power LED packages for different applications. There is single large die package like OSLON from OSRAM, XLamp XM-L3 from CREE, and K2 from Lumileds. Multiple large die size in a package like OSTAR series of OSRAM. Small/medium-sized die array in a package from Luminus and OSRAM. And single or multi very large die in a single package of Luminus or OSRAM.

The emergence of high-power LEDs is mainly due to the inability of low-power or mid-power packages to handle heat dissipation. For example, if the input power of an LED is 1 W, with a chip of internal efficiency of 50%, the heat generated in the LED will be roughly 0.5 W. This heat has to be dissipated from the chip efficiently and quickly. Failing to do so, the chip will have a significant impact on brightness and package reliability, and its lifetime will deteriorate. In many cases, the LED failed spontaneously. Figure 1.12 shows the high-power LED PLCC package thermal management design. It has big heat slug (sink) that allows the heat from chip to drain into the heat sink. The big heat slug also plays a role as a good thermal capacitance reservoir.

Beside the PLCC package, there is also ceramic package in high-power LED products. In Figure 1.13, there is an example from OSRAM that shows the OSLON®Square ceramic package. The substrate of this package is usually aluminum nitride, but there are also some cases made of silicon carbide. The thermal conductivity of silicon carbide is higher than that of aluminium nitride. This is ideal for high-power LED. However, the price of silicon carbide is relatively higher than aluminium nitride.

The die is attached to this ceramic substrate. On top of the die, a phosphor layer was attached. It was encapsulated using clear silicone. This encapsulant usually forms an optic to collimate the light at a certain viewing angle.

The LED high-power packages evolved mainly for automotive applications in the mid-1990s, where high-brightness LED makes a significant market differentiation in terms of aesthetic values, clarity, and elegance that charms end users. Luxury car makers such as Mercedes, BMW, Audi, and others are capitalizing on these values

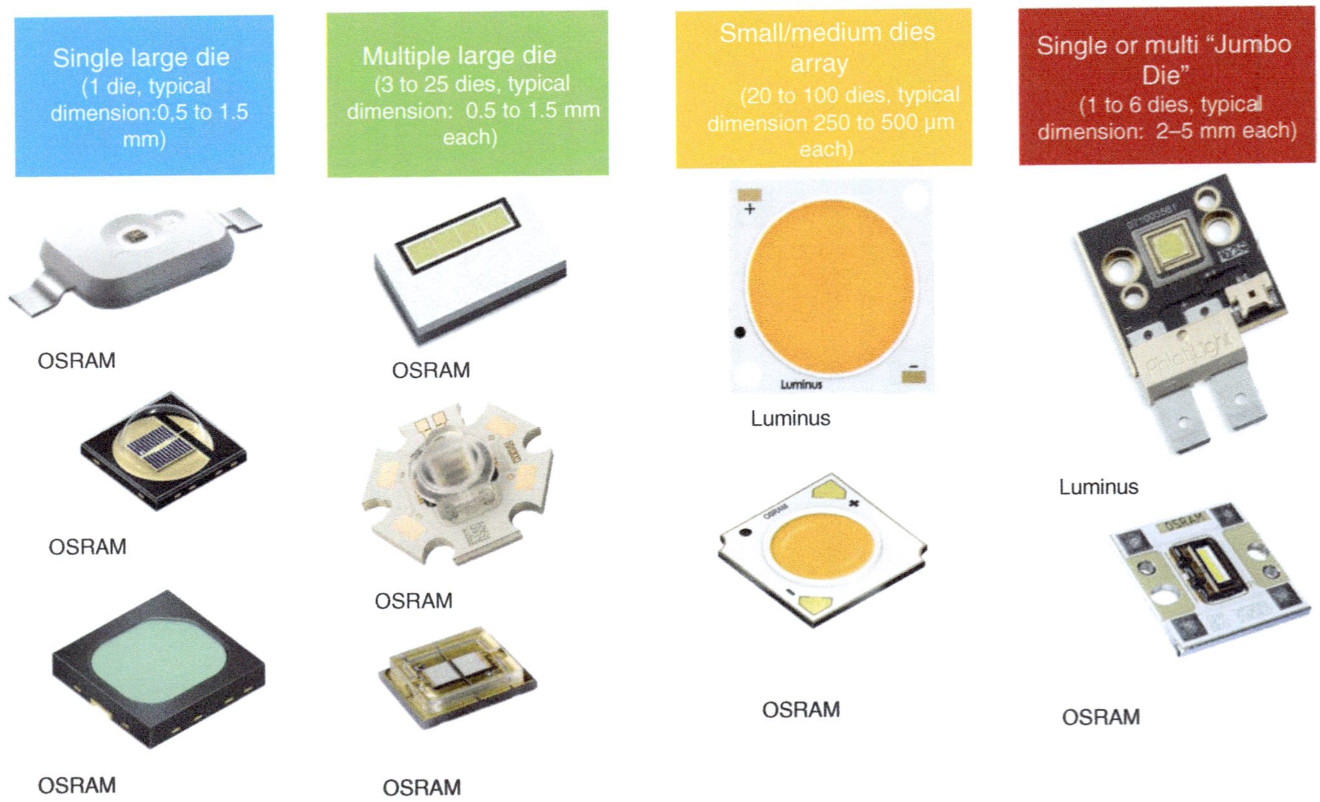

Figure 1.11 High-power LED packages with a variety of solutions. Source: Courtesy of ams OSRAM GmbH and Courtesy of Luminus, Inc.

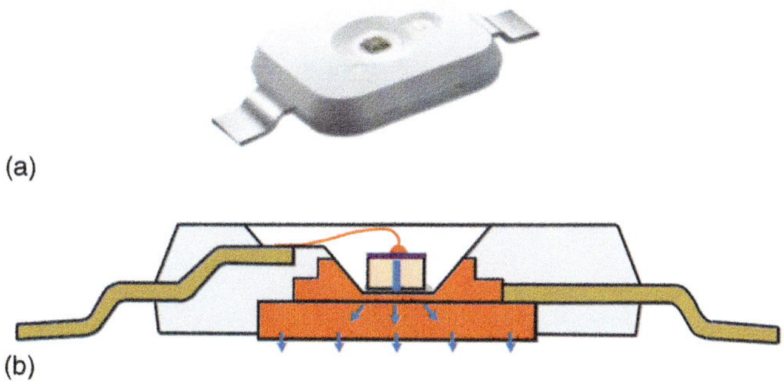

Figure 1.12 High-power LED PLCC package thermal management design: (a) Golden Dragon PLCC package. (b) Cross-sectional view of Golden Dragon PLCC package. Source: Courtesy of ams OSRAM GmbH.

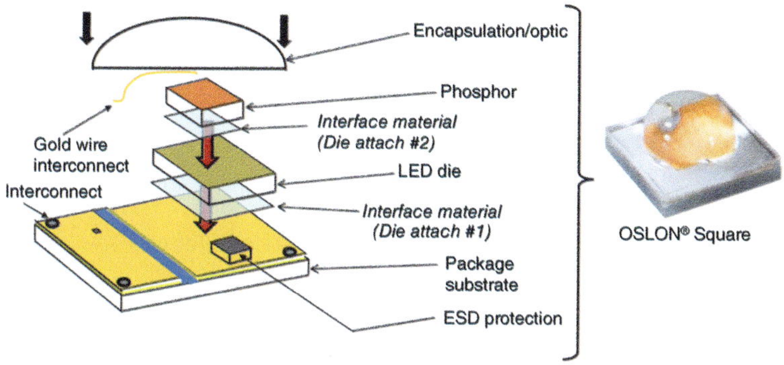

Figure 1.13 High-power LED package construction. Source: Courtesy of ams OSRAM Group.

to increase their sales. As a result, many of these large companies collaborated on research projects to develop new LED products that add value to end users. Some of the high-power packages that have evolved over the years are OSRAM Golden Dragon, Luxeon K2, Cree XLamp, Lumileds Rebel, OSRAM OSLON Square, OSRAM OSTAR, OSRAM OSLON Pure CSP, and SiP. Each is used for a specific application in the car, which will be further elaborated in a Chapter 2 and 4.

1.5 Summary

Light is one of the most essential elements for all living things. Humans mastered the science, technology, and engineering of artificial light over a period of time. This mastery has changed their lives and the course of evolution. The artificial light itself has evolved. The energy conversion efficiency of artificial lighting has changed from less than 1% to almost 80% in today's LED. The LED light has revolutionized human civilization. The LED application now covers almost every aspect of human life. The LED packaging plays a very important role to fit LEDs into all the applications that necessitate human wellbeing. It is the bridge between the LED chip and LED applications. The LED industry itself is still evolving. The LED's brightness increased as the need for high brightness increased. The LED packages evolved to support high brightness, where the package thermal management design was further perfected. The evolution further continues from single-component LED to a combined LED, sensor, IC, and passive component to a system packaging or system in a chip.

References

1 Alperson-Afil, N. (2008). Continual fire-making by hominins at Gesher Benot Ya'aqov, Israel. *Quaternary Science Reviews* 27 (17–18): 1733–1739.

2 Walter, C. (2015). The first artists. In: *National Geographic Magazine*, 33–57. Washington, DC: National Georgraphic Society.

3 Nordhaus, W.D. (1996). Do real-output and real-wage measures capture reality? The history of lighting suggests not. In: *The Economics of New Goods* (ed. W.D. Nordhaus), 27–70. University of Chicago Press.

4 Zukauskas, A., Shur, M.S., and Gaska, R. (2002). *Introduction to Solid-State Lighting*. New York: Wiley.

5 Alferov, Z.I. (2013). The semiconductor revolution in the 20th century. *Russian Chemical Reviews* 82 (7): 587.

6 Burton, F.D. (2011). *Fire: The Spark that Ignited Human Evolution*. UNM Press.

7 Holonyak, N. Jr., and Bevacqua, S. (1962). Coherent (visible) light emission from $Ga(As_{1-x}P_x)$ junctions. *Applied Physics Letters* 1 (4): 82–83.

8 Grimmeiss, H.G. and Allen, J.W. (2006). Light emitting diodes – how it started. *Journal of Non-Crystalline Solids* 352 (9–20): 871–880.

9 Yam, F.K. and Hassan, Z. (2005). Innovative advances in LED technology. *Microelectronics Journal* 36 (2): 129–137.

10 Schubert, F.E. (2006). *Light Emitting Diode*, vol. 2. New York, USA: Cambridge University Press.

11 Nakamura, S. (1991). GaN growth using GaN buffer layer. *Japanese Journal of Applied Physics* 30 (10A): L1705.

12 Nakamura, S., Pearton, S., and Fasol, G. (2013). *The Blue Laser Diode: The Complete Story*. Berlin, Germany: Springer-Verlag Berlin Heidelberg.

13 Ehrentraut, D., Meissner, E., and Bockowski, M. (2010). *Technology of Gallium Nitride Crystal Growth*, vol. 133. Springer Science & Business Media.

14 Wierer, J.J., David, A., and Megens, M.M. (2009). III-Nitride photonic-crystal light-emitting diodes with high extraction efficiency. *Nature Photonics* 3 (3): 163–169.

15 Wright, M. (2014). Research projects five years of growth for packaged LEDs and SSL. *LEDs Magazine* (22 April), 1–18. http://www.ledsmagazine.com/articles/print/volume-11/issue-4/features/markets/research-projects-five-years-of-growth-for-packaged-leds-and-ssl.html.

16 Zhu, D. and Humphreys, C.J. (2016). Solid-state lighting based on light emitting diode technology. In: *Optics in Our Time* (ed. M.D. Al-Amri, M. El-Gomati, and M.S. Zubairy), 87–118. Cham: Springer International Publishing.

2

Fundamentals of LED Packaging Technology

Packaging technologies for LED have received great attention lately. It is important to have appropriate packaging to realize the full potential of an LED chip. A bare LED chip cannot be used directly without a package due to lack of protection to the chip. An LED chip needs packaging because it not only ensures good performance of LED devices by enhancing reliability and optical characteristics but also is used to control and adjust the final performance. As a prerequisite of LED use, LED packaging plays an important role in determining the final optical and thermal performance of LED devices. In summary, there are five key functions of LED packaging:

a. Effective light extraction.
b. Package design and encapsulation.
c. Thermal management.
d. Electrical interconnection.
e. LED light conversion.

In Sections 2.1–2.6, these key functions of the LED package will be explained in detail. This includes LED science as the fundamental of the LED light.

2.1 Effective Light Extraction

There are many ways of extracting light effectively from LED chip. In this section, these techniques to extract light will be explained in detail. This includes those at chip level and package level.

It is well understood in the LED industry that the cost to increase the chip brightness is far greater than the effort for saving every photon that is lost either in the chip itself or in the LED packaging. As a result of this, much research has been done to improve the efficacy of the LED. Before we deep dive into LED efficacy, one has to understand the difference between LED luminous efficacy and efficiency.

Efficacy of a light source is how well it turns input power into the desired output, which is lumens. Luminous **efficacy** is a measure of how well a light source produces visible light. It is the ratio of luminous flux to power, measured in lumens per watt in the International System of Units [1]. The luminous efficacy of a light source is its generated luminous flux divided either by its radiant flux or by its electrical

LED Packaging Technologies: Design, Manufacture, and Applications, First Edition.
Luruthudass Annaniah, Mohamed Salleh M. Saheed, and Rajan Jose.

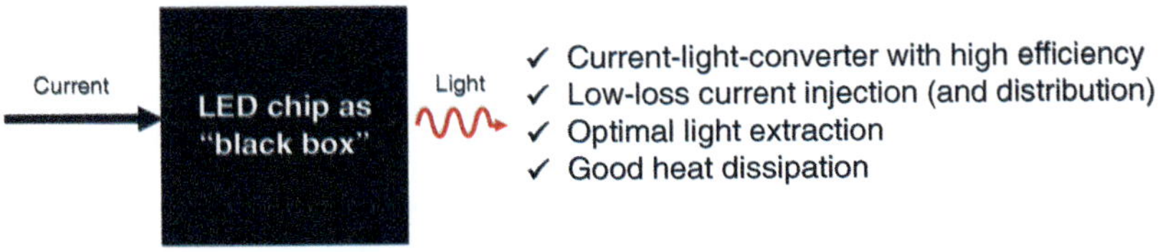

Figure 2.1 LED converts electrical energy into light. Source: Courtesy of ams OSRAM GmbH.

power consumption. In both cases, one obtains units of lumen per watt (lm/W), but the meaning is of course different. The more common definition is the luminous flux divided by the electrical power consumption of a light source. For example, a 60-W light bulb (incandescent lamp) may emit 960 lm, which leads to a luminous efficacy of 16 lm/W. The same luminous flux could be generated with a lamp based on light-emitting diodes (LEDs) with power consumption of 10 W, which comes to 96 lm/W. Hence, the LED-based lamp is much more efficacious.

LED luminous **efficiency**, on the other hand, is the actual percentage of power, which comes out as photons. The luminous efficiency of a light source is generally defined as its luminous efficacy divided by the maximum possible value of the efficacy. That raises the question of what exactly is meant by "maximum possible." One can take the ideal value of 670 lm/W as achieved for an ideally efficient light source meeting at 560 nm. In that case, a light bulb with 16 lm/W will have a luminous efficiency of 16/670 = 2.4%. On the other hand, if we take an LED-based lamp at 96 lm/W, will have lumen efficiency of 96/670 = 14.3%. Even an ideally energy-efficient white light source could then never reach 100% luminous efficiency, since the efficacy is necessarily reduced for the red and blue spectral components, for example.

In order to understand the LED's energy conversion from electrical to light, one has to assume the LED as a black box, as illustrated in Figure 2.1. When a certain amount of electrical energy is injected into an LED, the LED will convert most of the electrical energy into light in a process called radiative recombination, where electrons and holes combine to produce light. There will be some energy losses during current distribution as an electrical loss. Some of the energy will be converted to heat due to the internal efficiency of the LED chip. This energy loss in the chip is due to non-radiative recombination process. It is also called "phonon" – lattice vibration generating heat. Generation of phonons increases with heat. Some of the light was absorbed by the chip and some by the package itself.

The efficiency of an LED can be calculated using the Eq. (2.1) [2]:

$$\eta_{wp} = \eta_{int} \cdot \eta_{el.} \cdot \eta_{extr} \cdot \eta_{p} \tag{2.1}$$

η_{wp} = wallplug efficiency of an LED

η_{int} = internal quantum efficiency of the chip

$\eta_{el.}$ = electrical losses

η_{extr} = light extraction efficiency from the chip

η_{p} = package efficiency (light extraction, thermal management, light conversion)

The **wall-plug efficiency (η_{wp})** of an LED is defined as the fraction of the input power (IV) that is converted to optical output power. Wall-plug efficiency is the ratio of the radiant flux (i.e. the total radiometric optical output power of the device, measured in watts) and the electrical input power, i.e. the efficiency of converting electrical to optical power.

Internal quantum efficiency (η_{int}), is the proportion of all electron–hole recombinations in the active region that are radiative, producing photons. It is the **efficiency of the LED** chip. The energy supplied to chip, some are losses during the current flow from the source to the active region. This energy loss is called **Electrical loss ($\eta_{el.}$)**.

Not all photons from the active region are escaping from the chip, some are absorbed by substrate and chip materials. The efficiency of the light escape is called **light extraction efficiency (η_{extr})**.

The photons that escape from the chip are sometimes absorbed by the package. The package plays an important role in enhancing the efficiency of the LED. It manages the thermals of the chip. Higher the temperature, less efficient the chip is. Some LEDs require conversion; hence, the converter conversion affects the overall LED efficiency. All these influences directly affect the **LED package efficiency (η_p)**.

We have discussed LED's efficiency and LED light conversion from electrical energy into light. However, what is the science behind the LED light that converts electrical energy into optical energy? To answer this question, one has to understand the theory of light conversion in LEDs. In Section 2.1.1, the theory will be briefly explained.

2.1.1 Theory of Light Conversion in LED

Most of the LEDs in the world today are constructed using semiconductor material alloys that are based on groups III–V of the Periodic Table [1, 3].

The most commonly used materials are gallium arsenide (GaAs), indium gallium aluminum phosphate (InGaAlP), aluminum indium gallium arsenide (AlInGaAs), gallium arsenide phosphite (GaAsP), and gallium phosphide (GaP) and gallium nitride (GaN) [1, 4]. Each of these material systems emits different wavelengths that are directly related to their bandgap energy. Light is produced when an LED is biased at a certain voltage and current. Here, an electron crosses the junction from n- to p-type material, where the electron–hole recombines [5, 6]. This recombination process produces photons (light) at IR or visible wavelengths in a process called electroluminescence, as illustrated in Figure 2.2 [1].

In a more elaborate explanation, an LED works as per the illustration in Figure 2.3a, which shows the p–n junction situation at forward biasing conditions of a simple LED. Figure 2.3b shows the band diagram of the heterostructure of an $Al_x(Ga_{1-x})In_{0.48}GaP$ LED with quantum wells inside the p–n-junction. When a forward voltage is applied to the structure (positive to p-side layer and negative to n-side layer), the electrons are injected from the n-side layer into the p-side layer and holes from the p-side layer into the n-side layer, as shown in Figure 2.3a. These injected carriers are called minority carriers because they contain a relatively small

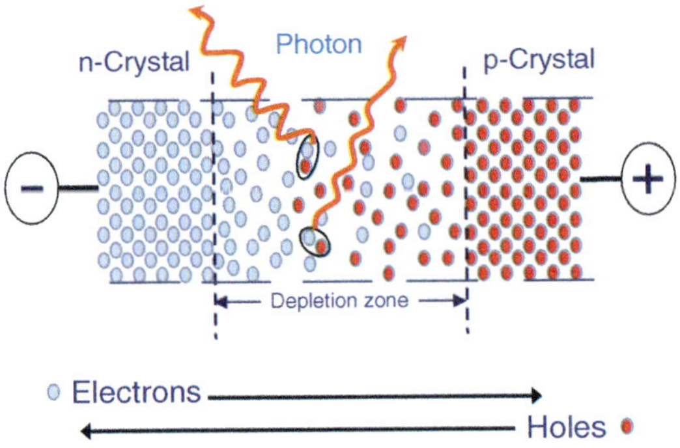

Figure 2.2 LED p–n junction in forward bias. Recombination occurs in the depletion region that produces photons (light). Source: Adapted from Schubert et al. [1].

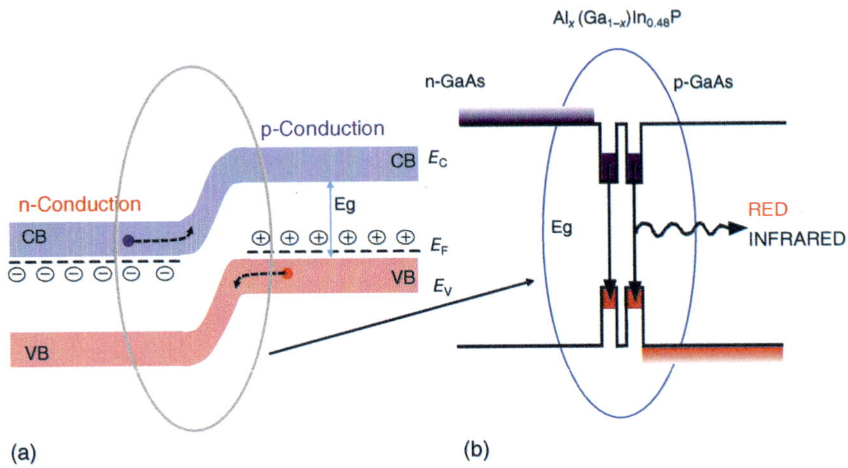

Figure 2.3 p–n junction under the forward bias conditions, minority carriers diffuse into the neutral regions, where they recombine. (a) Simple LED with p–n junction. (b) AlInGaP LED with quantum wells inside the p-n-junction. Source: Schubert et al. [1]; World Wide Web [7].

number of electrons surrounded by a large number of holes on the p-side and vice versa on the n-side. These electrons and holes can be in the same physical space, but they are separated by the energy gap or bandgap, E_g as shown in Figure 2.3b [1, 8].

In a very simple analogy, a positively charged hole is nothing else but a missing electron in the crystal lattice. Both electrons and holes can move freely through the crystal lattice. By applying a positive voltage or forward voltage (V_F) to the p-side electrode, electrons will diffuse from the n-side to the p-side of the conduction band. Similarly, holes will diffuse from the n-side of the valence band.

A positively charged hole can attract a negatively charged electron, and the electron can recombine with the hole and release photons (light). However, this process has to obey two fundamental laws of physics that are energy and momentum conservation [3, 9].

The law of energy conservation can be equated with quantum energy law, which is $E_g = h\nu$. This process results in the conversion of an injected electron or hole into photons as long as the energy gap is in the range of 1.9 eV (red) to 3.0 eV (violet) for visible light and 1.4–1.8 eV for IR light [10]. Direct bandgap materials, such as GaAs, GaAlAs, GaInN, and GaAlIP, have injected electrons that readily recombine with holes by emitting infrared or visible light with a wavelength depending on the bandgap energy E_g [10]. GaN-based material systems have a bandgap energy of approximately 3.47 eV. GaP has a bandgap energy of 2.5 eV, and GaAs, on the other hand, has a bandgap energy of 1.52 eV [3, 10]. However, electrons also have a chance to recombine without emitting light. To recombine radiatively, the electron (or hole) must find a hole (or electron) with the exact opposite momentum to meet the law of momentum conservation.

Radiative recombination requires momentum conservation, and recombination probability of an electron is proportional to the number of holes available at the same momentum. The recombination probability decreases with the increased temperature. The recombination process will take some time. During this time delay, the electron (or hole) has a finite probability of dropping into an electron (or hole) trap such as a crystal defect or crack [6]. While being trapped, the electron (or hole) will eventually recombine with a hole (or electron). Instead of generating a photon, this recombination process by emitting multiple phonons or lattice vibrations (heat). This is not desirable in the LED industry [1].

Considering these two recombination paths, radiative and non-radiative, the efficiency of the recombination process can be described by a simple equation as shown below [9]:

$$\eta_{\text{int}} = t_{\text{n}}/(t_{\text{n}} + t_{\text{r}}) \tag{2.2}$$

where η_{int} is the internal quantum efficiency, t_{n} is the mean time to recombine non-radiatively, and t_{r} is the mean time to recombine radiatively. The best case is a very small $t_{\text{r}} < t_{\text{n}}$, then $\eta_{\text{int}} \sim 1$. This means the efficiency of the electrons that recombined radiatively at quantum is close to 100%. However, the actual condition of LED efficiency is still relatively lower than this. The non-radiative combination occurs due to several physical mechanisms. These include the defects in the epi-layer, for example, cracks, dislocations, foreign atoms, and any complexes of the defects [10]. All defects have energy-level structures that are different from the substitutional semiconductor atoms. It is quite common for such defects to form one or several energy levels within the bandgap of the semiconductor [9]. Many factors are causing these defects. Minimizing these defects, such as cracks and dislocations, is the key to gaining superior LED performance.

2.1.2 Light Extraction Based on Chip Technology

Despite the high internal quantum efficiency (IQE) of the chip, if the photon is not outcoupled effectively, the brightness of the LED will not be high. It is important to know what light extraction technology is available in an LED chip and how it works.

The growing demand for blue light LEDs has also prompted for devices with a maximal external quantum efficiency (EQE), which is determined by both internal quantum efficiency (IQE) and light extraction efficiency (LEE). With

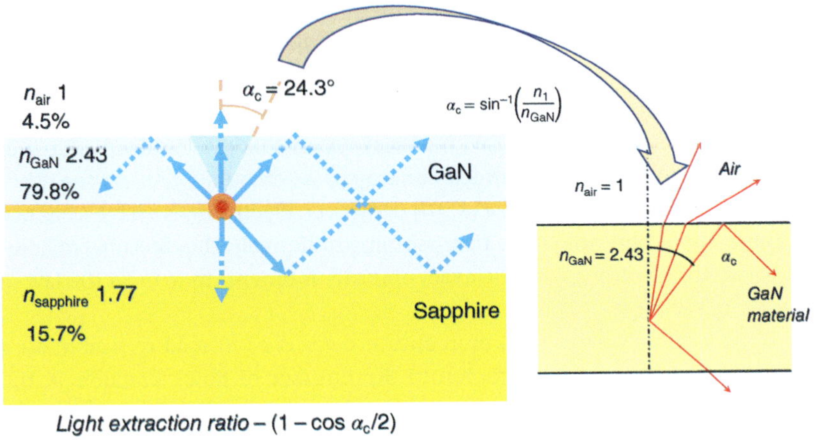

Figure 2.4 Light paths in GaN/sapphire chip.

the rapid and massive improvements in growth techniques, epitaxial structures, and crystal quality, the IQE has been greatly enhanced to more than 80% [11]. However, the extremely low extraction efficiency (<10%) is still one of the major bottlenecks restricting the performance of LEDs, attributed to the absorption of substrate, current spreading layer, ohmic contacts, and bonding wire, as well as the main challenge of total internal reflection (TIR), thus implying that there is still plenty of room for improving the LEE. In the following parts, the influence of TIR is explained. Numerous approaches aiming to extract optically guided light from devices and suppress TIR are highlighted, including surface roughening, buried micro-reflectors, geometrical shaping, and photonic crystal. These methods rely on the formation of nonparallel surfaces to minimize reflections and reduce reabsorption loss, albeit at different dimensional scales.

One of the limiting factors in LED efficiency is the light trapping in the LED chip. This is mainly caused by internal reflection, as illustrated in Figure 2.4. The angle of TIR defines the light escape cone. The angle of TIR between a GaN chip and air, α_c is 24.3°, which is estimated using the **Snail law** as shown in Eq. (2.3). This means any light emitted beyond 24.3° will be reflected back into the chip.

$$\alpha_c = \sin^{-1}\left(\frac{n_{air}}{n_{GaN}}\right) \tag{2.3}$$

n_{air} = index of refraction of air = 1

n_{GaN} = index of refraction of GaN = 2.43

$n_{sapphire}$ = index of refraction of sapphire = 1.77

By calculating the ratio of solid angle escape cone to solid angle emission, one can estimate the percentage of light escaping out of the chip. The escape ratio is given by Eq. (2.4).

$$(1 - \cos(\alpha_c))/2 \tag{2.4}$$

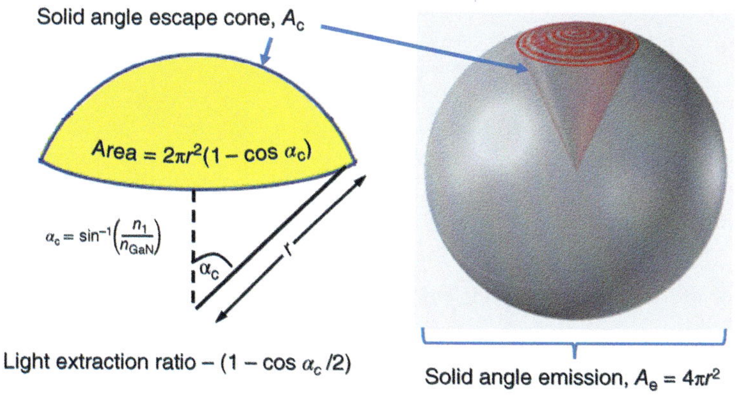

Figure 2.5 Solid angle escape cone. Dark line indicates the critical angle for total internal reflection.

This is derived by taking the ratio of light solid angle escape cone, A_c to the solid angle emission, A_e. The solid angle escape is shown in Figure 2.5.

$$A_c/A_e = 2\pi r^2(1 - \cos(\alpha_c))/4\pi r^2$$

By using the Eq. (2.4) light escape from the top of smooth surface sapphire chip is roughly 4.5%.

A smooth surface chip, GaN on the sapphire, only 4.5% of the light emitted from quantum well is emitted from the top surface of the chip of an unencapsulated LED. Nearly half of the light emitted from the chip quantum well radiated downward toward the substrate [1]. The GaN/sapphire substrate boundary, the critical angle is 46.7°, and about 15.7% of emitted light is coupled into the substrate and lost. The remainder of the light, roughly 79.8%, remains trapped inside the GaN layer. Some of the trapped light reflected back into the epitaxial layer is scattered from epitaxial defect sites to an angle lying between the allowed extraction cone. Some of the recycled light is also emitted from the edge of the LED chip, so the actual light extracted from the chip is relatively higher than 4.5% [12].

Using the same Snail law, we can now estimate the TIR of the chip and package interface. Let's say the chip is encapsulated with silicone. A low refractive index (RI) silicone has an index of refraction of 1.46. The TIR is 38.9°. Using Eq. (2.4), light escape from the top of the smooth surface chip to the silicone encapsulant package will be roughly 31.9%. This is relatively high compared to air. This means more light can escape from the chip structure through package encapsulant into free space. The light extraction is enhanced. Hence, this is the reason, the packaging is important for enhanced light extraction. How can we enhance the chip light extraction? In Sections 2.1.2.1–2.1.2.3, the light extraction will be further enlightened.

2.1.2.1 Chip Surface Roughing

A popular, cost-effective, and practical approach to high extraction is to roughen the surface of the LED chip, as illustrated in Figure 2.6a. Common roughening techniques, including photo-electrochemical chemical etching and wet etching, are

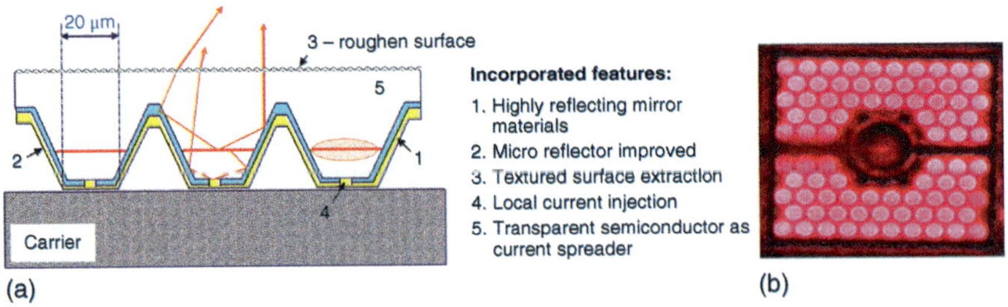

Figure 2.6 Buried micro-reflectors showing emission regions of the LED. Source: Courtesy of ams OSRAM GmbH.

capable of developing high-density, randomly oriented miniature facets/features on the LED surface [2]. The processed surface can randomize the path of trapping and significantly increase the probability of light striking the boundary at an angle close to normal. Surface roughing technique has proven to enhance the light output power and effectively scatter the trapped light outside the LED chip.

2.1.2.2 Buried Micro-Reflectors Chip

When emitted light incidents upon the boundary at an angle greater than critical angle; it suffers TIR and becomes laterally guided modes, which can either be reabsorbed or escaped from the edges of LED. To increase the chances for light to be extracted into free space before reabsorption, buried micro-reflectors coupled with roughened chip surface, as illustrated in Figure 2.6a,b, provide additional photon escape pathways. The light from the MQW is reflected directionally to free space vertically.

2.1.2.3 Chip Geometrical Shaping and Type

For a conventional LED chip with rectangle geometry as illustrated in Figure 2.7a, a light ray reflected from one face is likely to hit another parallel facet and bounce around inside LED chip until it is reabsorbed. One way to overcome this problem is to change the shape of the LED chip by creating beveled sidewalls such that the facet pairs are no longer parallel and possibly alter the propagation direction of reflected light. Wang et al. demonstrated the light output power of nitride-based LED chips with polygonal geometry, significantly improved the light extraction compared to the common rectangle-shaped LED chip. This is illustrated in Figure 2.7b, where the various polygonal LEDs shaped the light extraction improved significantly compared to the rectangle chip [13].

We have known that the chip's geometrical shape plays an important role in light extraction. Having known the light extraction at the chip level, it is important to know how the light extraction is at the package level. The chip type used for the LED plays an important role in the LED packaging design. There are many types of LED chips in use today. However, they can be categorized into three main types. There are Thinfilm, Sapphire, and Flip chip. Thinfilm chip and Sapphire chip have different radiation escape cones. The Thinfilm chip is those lights that are emitted vertically from chip top as illustrated in Figure 2.8a. On the other hand, the Sapphire

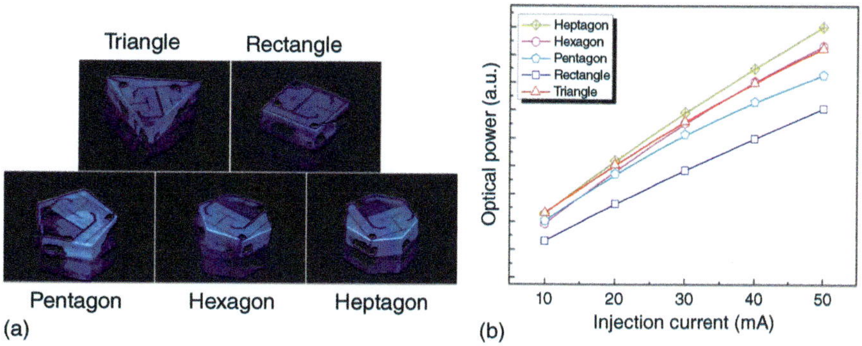

Figure 2.7 (a) Optical microscopy images of various LED chips with various geometrical shapes. (b) Measured light out power vs injection current (*L*–*I*) for laser micromachined polygonal LEDs. Source: Reproduced with permission from AIP Publishing LLC.

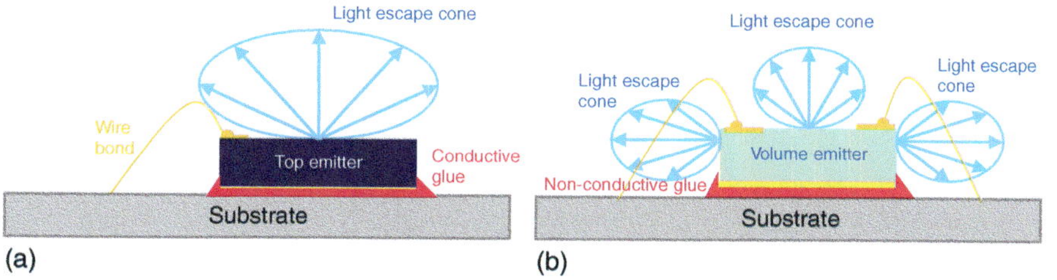

Figure 2.8 Light emission of the standard Top Emitter chip vs Volume Emitter chip. (a) Thinfilm chip. (b) Volume Emitter (Sapphire) chip. Source: Courtesy of ams OSRAM GmbH.

chip or Volume Emitter has light escape cone from the top and the four sides of the chip, as shown in Figure 2.8b. The standard Thinfilm and Volume Emitter chips have wire bonding that connects the electrical connections from the top of the die. This wire bond has some impact on the light extraction as the wire blocks a certain amount of light from the chip. This is where the Flip chip comes in handy to save some photons from the chip to improve the LEE. Flip chip has no wire connection.

In the GaN-based LED chip, the sapphire substrate is commonly used. It has lattice constant close to that of GaN, which is ideal for growing GaN epi on it. Mechanically, sapphire is very stable. However, it is a nonconductive material. Hence, electrical contact has to be designed either on top for the standard volumetric chip or at the bottom for the Flip chip.

The Flip chip radiation pattern, as illustrated in Figure 2.9, has light escape cone quite similar to the Thinfilm and Volume Emitter chip except for no wire-bonding electrical connection from the top of the chip; on the other hand, the electrical connections are from the bottom of the chip.

Knowing the light escape cone of the chip, one can design the package that fits to optimize the light extraction. Volume Emitter requires good reflectors or reflective encapsulation material like titanium oxide (TiO) to enhance light outcoupling, as illustrated in Figure 2.10c. On the other hand, Top Emitter chip, as shown in Figure 2.10a, doesn't necessarily require elaborate reflectors. However, having a

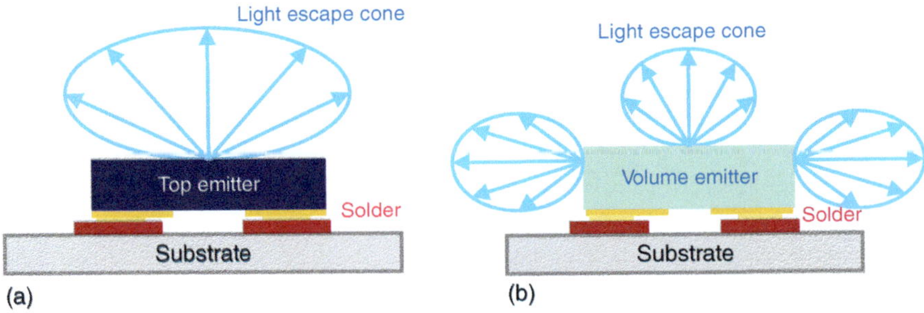

Figure 2.9 Light emission of the Flip chip Thinfilm vs Volume Emitter chip. (a) Flip chip–Top emitter. (b) Flip chip–Volume emitter. Source: Courtesy of ams OSRAM Group.

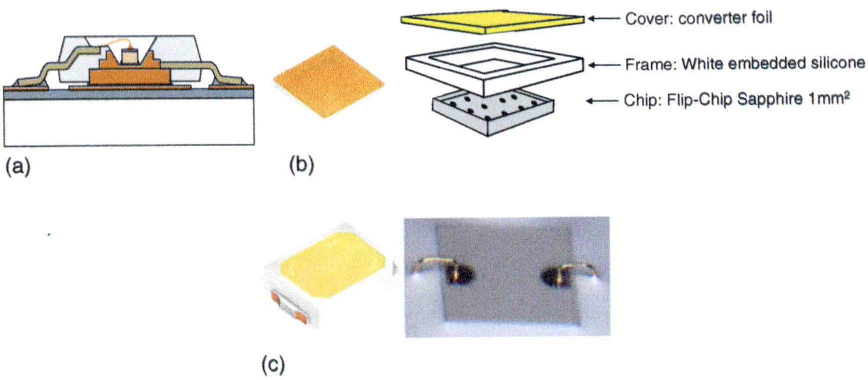

Figure 2.10 LED packaging that can be used for Thinfilm, Flip chip, and Volume Emitter chip. (a) Top Emitting package. (b) Flip chip sapphire package. (c) Volume emitter (sapphire chip) package. TiO_2 casted surround the sapphire chip before the encapsulation process. Source: Courtesy of ams OSRAM GmbH.

reflector does help to maximize light extraction even for Top Emitter, as some light is reflected back due to some light TIR of in-between encapsulant and air. Figure 2.10 shows a few varieties of the package that can be used for Top Emitting, Flip chip, and Volume Emitter (sapphire) chip types, respectively. The electrical connection comes from the wire bonding on top of the Thinfilm and Volume Emitter chip. Thinfilm chip uses a conductive glue at the bottom of the chip to connect it to the substrate. The sapphire chip, on the other hand, uses nonconductive glue and two wires for electrical connection. The Flip chip is soldered directly to the substrate. To improve the light extraction, some Flip chip-based packages have a white frame that acts as a reflector to maximize light outcoupling. as illustrated in Figure 2.10b.

2.1.3 Light Extraction Based on High Reflective Packaging Material

2.1.3.1 Leadframe Plating Surface Influence

In leadframe base, PLCC, and QFN packages, the leadframe reflectivity is crucial for maximizing the light extraction. The surface finishing of the leadframe plays an

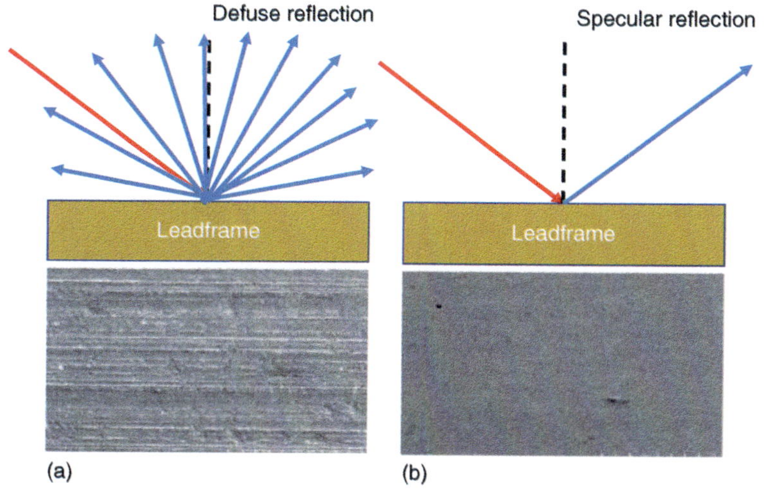

Figure 2.11 The lead frame plating surface finishing and light reflection. (a) Defuse surface and light reflection. (b) Specular surface and light reflection. Source: Courtesy of ams OSRAM GmbH.

important role in reflectivity [14]. Figure 2.11 shows the surface finish of the leadframe. Figure 2.11a shows defused surface, and Figure 2.11b shows specular surface. The specular surface reflects better than the defused surface, hence improving LEE. To get a specular surface, the surface finish has to be glossy. This can be obtained by using silver brightener in the plating bath. The plating thickness of the underlying layer (Ni, Cu) may also contribute to the smoothening of base material.

2.1.3.2 Housing Material Reflectivity

As we know, the light escape cone of the various LED chips shows light scatter in various patterns depending on the chip type. It is crucial to have highly reflective housing material to extract the lights effectively. This was confirmed by Kashiwao et al. in their investigation on the optimization of SMT LED packaging on LEE [14]. They use ray-tracing simulation to confirm the influence of surface reflection of the leadframe silver plating, encapsulation material, and the optical design of the SMT LED package. It is found that the optical properties of the encapsulation material, i.e. the housing material and leadframe silver plating, significantly influence the light extraction and, hence the package efficiency. The cavity angle formed by the cavity wall is also important to the optical design and significantly influences light extraction and package efficiency.

The PLCC package is usually molded using the injection molding technique. The material used for the package is usually thermoplastic material. One example of the PLCC package is OSRAM TopLED® as illustrated in Figure 2.12.

2.1.3.3 Encapsulation Material Light Extraction Efficacy

In the LED industry, there are several encapsulation materials used. They are transparent epoxy materials, hybrid silicones, and silicone materials. The hybrid silicone

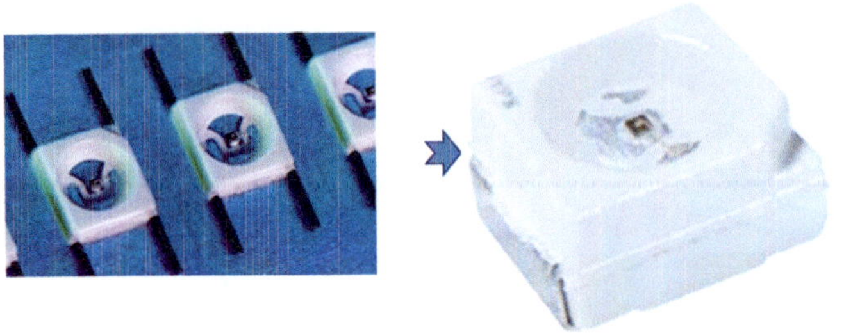

Figure 2.12 Injection-molded leadframe and OSRAM TopLED PLCC package. Source: Courtesy of ams OSRAM GmbH.

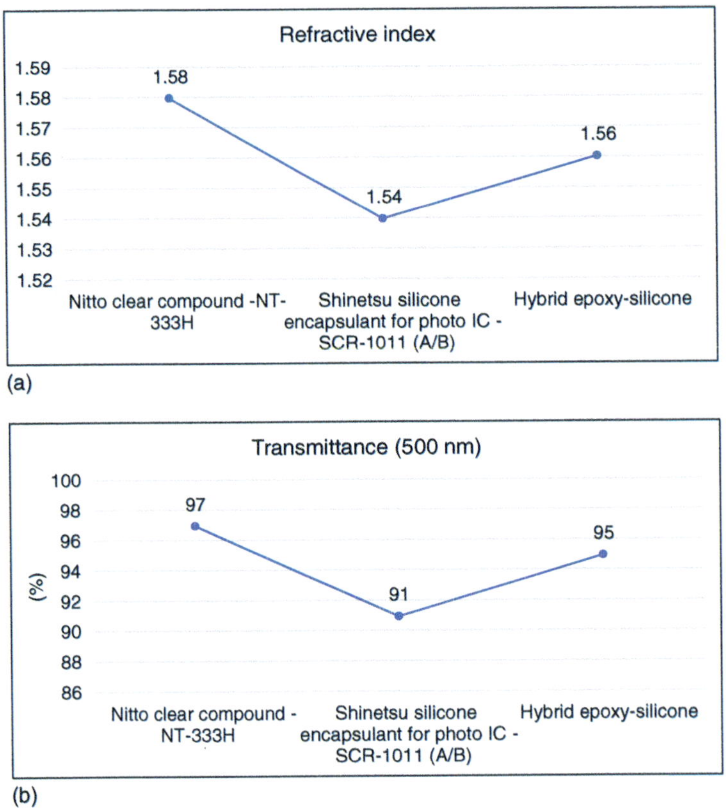

Figure 2.13 Optical properties of encapsulation material.

material is a material that was engineered by combining epoxy and silicone to get a material that has both advantages of epoxy and silicone [15]. The RI of encapsulation materials – epoxy, silicone, and hybrid as illustrated in Figure 2.13a show at 1.58, 1.54, and 1.56 relatively [16]. The higher the RI of the encapsulant material, the higher the light extraction will be.

Another important property of encapsulant material is its transmittance. The higher the transmittance of encapsulant material, the lesser the light absorption in the encapsulant material. Hence, more light will be outcoupled from the package into open space. Figure 2.13b shows the epoxy encapsulant has the highest transmittance at 97%, followed by hybrid at 95%, and lastly silicone at 91%. From this, we can see the epoxy encapsulant has better optical properties compared to silicone or hybrid. However, there are some weaknesses in an epoxy compound that silicone and hybrid are better off, for example, the Young's modulus, which is directly relevant to stress in the package and will affect the robustness of the LED.

2.1.4 Optical Interface Enhancing Light Extraction

Many applications of LED require a specific radiation pattern. Hence, the optical interface is important to ensure that the desired radiation pattern is achieved with minimum photon loss through absorption in the package and the optical element, while at the same time improving light extraction. There are several ways to meet the application of radiation pattern and light extraction. They are by lenses, package design, and chip design. In this section, we focus on the lens (optical interface). The lens can be directly molded on the chip or attached to a package.

A study done by Huang et al. shows a novel type of package called freeform-designed chip-scale package (FDCSP), which has ultra-high LEE. This unique package design is illustrated in Figure 2.14c.

For the backlight application, mainstream solutions are chip-scale package (CSP; Figure 2.14a) and surface-mount device package (SMD; Figure 2.14b). Comparing

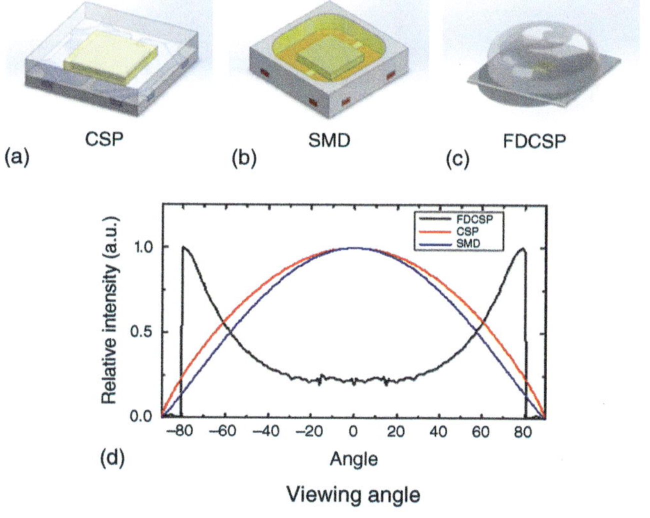

Figure 2.14 Schematic diagram of (a) CSP package, (b) SMD package, (c) FDCSP package and (d) viewing angle chart of CSP, SMD and FDCSP package. Source: Huang et al. © 2019, MDPI/Public Domain CC BY 4.0.

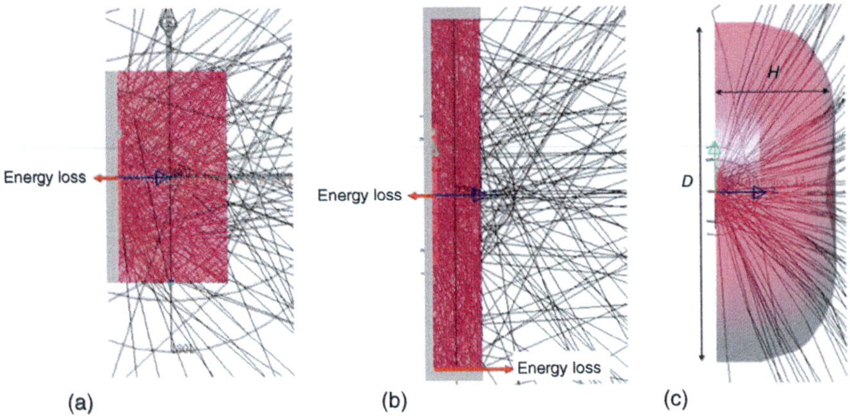

Figure 2.15 Light-tracing profiles of (a) CSP: 1.5 mm × 1.5 mm × 0.8 mm, (b) SMD: 2.5 mm × 2.5 mm × 0.35 mm, (c) FDCSP: D = 2.5 mm, H = D 0.8 mm. Source: Huang et al. © 2019, MDPI/Public Domain CC BY 4.0.

these two mainstream types of packages, the LEE of CSP, SMD, and FDCSP are 88%, 60%, and 96%, respectively, based on Huang et al.'s findings (Figure 2.14d). They used light tools to simulate the LEE of the packages. This light-tracing profile shown in Figure 2.15 clearly shows that the system efficiency is proportional to the number of rays coming out of the surface. Due to the design of the free-form surface, the FDCSP has a low proportion of the total beam reflection, resulting in minimal energy loss. As SMD is only a one-sided illumination and has no curved surface characteristics, the number of rays that the beam is reflected by and TIR in the package is very large, thus causing a large amount of energy loss [17]. Huang et al.'s finding confirms that the geometry of the package with curvature surface improves light extraction.

2.2 Package Design and Encapsulation Technology

As elaborated before, the package offers mechanical and environmental protection to the chip and provides good electrical connection with good heat dissipation to remove heat from the chip junction. At the same time, enable efficient light extraction from the chip. Package design and encapsulation go hand in hand as they are interrelated to each other.

2.2.1 Package Design

The package design of an LED is mostly defined by the intended customer applications. These packaging designs can be summarized into roughly seven categories that are interrelated to each other, as shown below:

a. Design for cost.
b. Design for reliability
c. Design for manufacturing.

d. Design for testing.
e. Design for the environment.
f. Design for assembly.
g. Design for effective light extraction.

All these categories are interrelated to each other. In a single package design, more than one of the above categories can be considered. Ultimately it has to fulfill the end customer's wish with the right cost, quality, and performance.

2.2.1.1 Design for Cost

Almost all the LEDs are designed specifically for an intended application where the cost of the LED per piece produced is the key winning factor in the product's success in the market. The LED package design must meet this criterion. For example, an LED is intended to be used as handphone flashlight. The target phone flashlight LED is expected to cost US$0.30 per piece. In the rule of thumb, the package cost for this LED is expected to be just a fraction of the overall cost of the overall LED cost. This means the package design must source materials that fit this design. The process must be simplified, i.e. designed for manufacturing, and must have a simple package design that meets the LED product reliability requirements of the intended application. Material selection for this case can use a QFN package, where the cost per piece is relatively cheap compared to other packaging and can be manufactured with excellent quality in the mass production process. The encapsulation material is recommended using epoxy with the converter material in a fully automated transfer molding process.

Cost-down pressure in the LED industry has driven tremendous innovation and cost-down measures to bring down the product cost. The packaging is one of the areas where a huge impact occurs in a wide variety of supporting industries. For example, the equipment suppliers for packaging have come up with a cheaper version of the equipment that reduces the capital cost. On the other hand, the material suppliers in the package are following the trend to bring down the cost as a new player comes in with very extractive prices for the materials. This as a whole helps the cost-down effort.

As a result all these cost-down pressures, the package design has to consider all these elements in their package design. Good package design is a package that can be manufactured in mass quantity with intended quality, high yield, and low cost. These can be achieved by understanding the customer's intended application and quality requirements and embedding them in the package design. For example, if a product requires a chip placement on a substrate with an accuracy of $\pm 10\,\mu m$, the equipment that is able to do this job with stable process capability of Cpk 1.67 will cost roughly US$500,000. If the placement accuracy increases to $\pm 50\,\mu m$, then the equipment that is able to do that job with stable process capability of Cpk 1.67 could cost only roughly US$100,000. It is a factor of 5 lower. It is important to understand the application of the product and package design so as not to over-engineer the package design to a level that requires high capital investment, which directly affects the overall cost of the LED package.

2.2.1.2 Design for Reliability

The reliability requirement of an LED product depends on the end customer application. The reliability requirement of an application in automotive product, for example, a headlamp is more stringent than that of a consumer application such as a printer in the office. The reliability test requirement of an automotive headlamp, for example, the temperature cycle test, the condition is at −40 °C to 125 °C. Hence, the constituent material in the LED package must be able to withstand a reliability test in the temperature cycle of −40 to 125 °C. This is, in contrast, less stringent compared to LED for an office printer, which is just −40 to 85 °C. In conclusion, the design of the automotive headlamp LED has to be robust.

Besides the temperature stability, some applications require chemical stability for specific application, for example, gas detection. Here, hermetically sealed package is required. Some of the examples include a ceramic package with a glass-sealed or metal-can package. Some LEDs are with short-wavelength, for example, UV LEDs, which badly affect the carbon-based packaging materials. It turns the encapsulation material brown and, to certain extent, weakens the package mechanically, where it easily cracks. For this application, a ceramic package with glass cap encapsulation is required. If it is for high-power application, the thermal management has to be good. Here, thermal via is needed to dissipate the heat out of the chip's active region in the most efficient way to avoid droop and maintain the long-term LED lifetime performance.

2.2.1.3 Design for Manufacturing

As mentioned in the earlier chapters, the cost of an LED is critical as the industry is going through aggressive cost-down pressure. One of the actions taken by many LED players is embarking on a design for manufacturing. It fosters manufacturing needs in the design of the package. This encompasses simplified package design, low cost and stable material performance, and robust process. Equipment selection that is easily able to handle the designed package. The process capabilities must have Cpk above 1.67, which provides a good confidence level for the manufacturing of the LED. An ideal LED manufacturing process has very low detection gate, i.e. visual inspection process for quality control. This again depends on the intended product applications. For automotive applications, the industry desired defects per million (dppm) less than 0.1 dppm. However, for some consumer markets, generally desired defect per million is roughly 1–5 dppm. Some of the automotive applications are for human safety, for example, for crash sensor applications. For this application, the market expectations are very stringent, process stability must be high, and very tight inspection control is required.

If the package design is not suitable for manufacturing, there will be more losses due to manufacturing defects. This eventually increases the cost of manufacturing.

2.2.1.4 Design for Testing

Today in the LED industry, the LEDs are commonly 100% tested before shipping to customers. Since it is 100% tested, testing can be a bottleneck process. If it is a bottleneck process, more investment is needed to debottleneck it by adding more equipment. To avoid such issue, this process must be highly efficient and effective. To support this, the package design has to be fit for testing. The location and size of the lead on the package have to be designed to ensure good electrical contact. Kelvin contact failure has to be very minimum. Kelvin contact failure units are mostly good units, but they failed due to bad contact between package lead and pogo pin. Commonly, Kelvin contact failure is controlled by less than 0.2%. However, this control varies from company to company. To minimize the Kelvin contact failure, one has to have a well-plated surface on the contact pin at the package. The lead-to-lead coplanarity variation has to be controlled at less than 0.1 mm. At the same time, the surface of the lead must be clean of contamination. Studies showed contamination on the surface of leads causes Kelvin contact failure. Figure 2.16 shows contaminated pogo pin and package lead that cause Kelvin contact failure.

The plating surface of the lead has to follow certain design rules so that it maintains certain mechanical properties that avoid lead plating surface pealing during touching the pogo pin contact with the lead. This metallic material will accumulate on the pogo pin surface and can cause Kelvin contact failures.

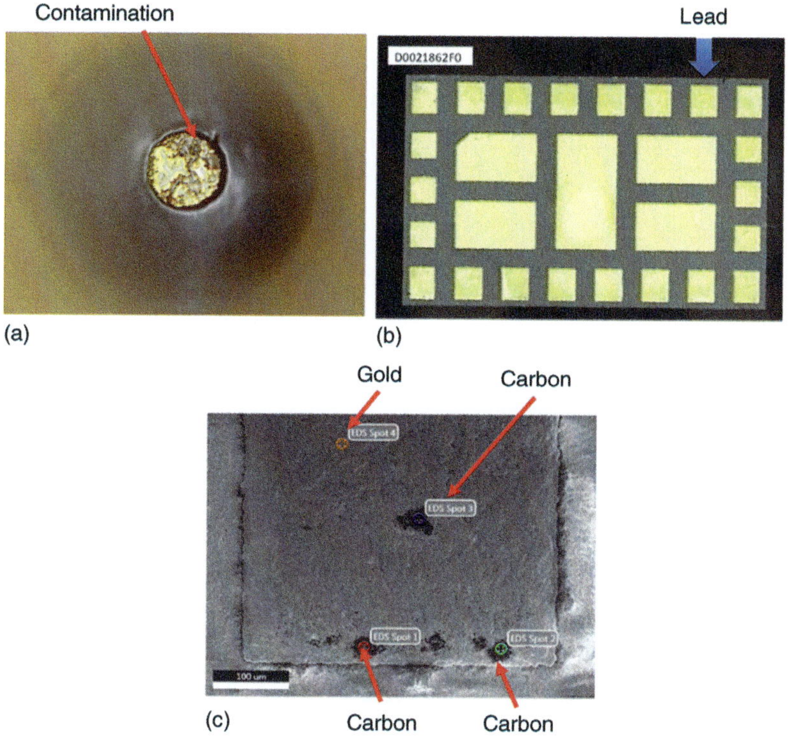

Figure 2.16 LED package external contact pin surface quality affecting testing. (a) Contaminated pogo pin surface. (b) Package external contact pin (lead). (c) EDX analysis indicate carbon residue on the lead. Source: Courtesy of ams OSRAM GmbH.

2.2.1.5 Design for Environment

The LED package design varies across the various LED applications. The environment that the LED is intended to be used in, plays an important role in the package material selection, and the encapsulation concept. An example of an LED used in the automotive industry as a gas detection sensor in the exhaust needs a package that is hermetically sealed and able to withstand the harsh exhaust environment. Here, a good package to withstand this environment will be a ceramic package with a glass cap sealed in an inert gas environment. On the other hand, for an LED intended for use in a printer that operates at room temperature conditions, a CSP or PLCC package design with epoxy or silicone encapsulation material is sufficient.

2.2.1.6 Design for Assembly at Second Level PCB Board

The LED package design must take into consideration of assembly at customer PCB board. The contact point between LED and customer PCB board is the LED leads. The anode and cathode lead spaces must have sufficient clearance to avoid solder bridge failure. Figure 2.17 shows an example of CSP LED lead contact design and

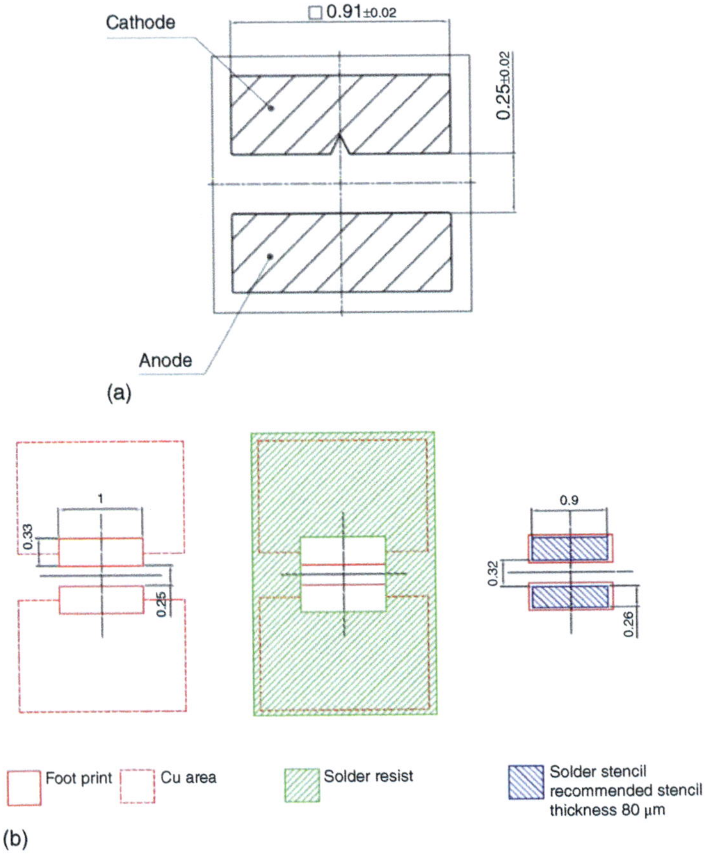

Figure 2.17 (a) CSP lead contacts and (b) recommended PCB solder pad design. Source: Courtesy of ams OSRAM GmbH.

the recommended PCB solder pad design. The lead coplanarity has been less than 0.1 mm. The plated surface of the lead has to be free of copper exposure. Exposed copper surfaces have poor solder wettability. This will cause dewetting, and hence, poor contact between LED and the PCB.

The thickness of Au, Ag or AuPd alloy plating on the lead must follow the design rule. A thin-plated lead can cause poor solder quality (solder void) at the LED lead – PCB board interface. This is a defect. In some high-power LED applications, the solder void is controlled at less than 10%. Hence, the package design must carefully consider the plating thickness at the contact leads.

2.2.1.7 Design for Effective Light Extraction

One of the key roles of a package is to extract light effectively. This has been elaborated in the earlier chapter. As mentioned earlier, every photon saved from the chip is a gain in brightness enhancement for the overall LED. Many types of designs can help to have a good out-coupling of the photon from the chip. A reflector in the package will converge the light in the intended direction. The packaging material must have good reflectivity. Usually, titanium oxide (TiO) or aluminum oxide (Al_2O_3) are used to have a good light reflection. The anode and cathode lead plating materials also absorb light. For short-wavelength, i.e. UV to blue, gold-plated surface is not so idle. Usually use Ag, but in some applications where Ag is not inert enough to withstand corrosion, gold was used. If gold were used, TiO would be used to cover the gold surface and improve the light reflectivity.

Certain application requires very specific radiation pattern. Here, the package can be designed with lens to provide specific radiation pattern that the application requires.

2.2.2 Encapsulation of LED

Encapsulation is one of the key packaging features to enable mechanical and environmental protection in the LED packaging and, at the same time, assist the efficient light extraction from the chip. There are many types of encapsulation. Following are some of the common encapsulation concepts:

a. Epoxy, silicone, and hybrid compound encapsulation.
b. Hermetic sealing encapsulation.
c. Epoxy-cap encapsulation.
d. Glass-cap encapsulation.

Each of the above encapsulation components has its own specific purpose, which is derived from the LED application.

2.2.2.1 Epoxy, Silicone, and Hybrid Compound Encapsulation

Epoxy compound encapsulation is one of the most common encapsulation methods for LED packaging. It has been in the industry for more than a half-century and is still in use today.

The advantage of epoxy encapsulation materials is that they are hard, and their coefficient of thermal expansion (CTE) is adapted to LED applications. It has excellent adhesion to housing materials and leadframe materials. It has a low diffusion coefficient for gases. Since it is a hard material with low coefficient for gases, it provides excellent protection to LEDs from chemical and mechanical stress exposure. Epoxy material is very cheap compared to silicone or hybrid compounds. This is ideal for low-cost LEDs. However, disadvantages of the epoxy compound are it is not suitable for prolonged use at high power short-wavelength rays like blue or UV light application as the short wavelength radiation affect the carbon molecule chain in the epoxy molecule. It breaks the carbon bond from the epoxy molecule. It is called "carbonization." The clear epoxy turns dark for prolonged exposure, as shown in Figure 2.18, which affects the light transmission and makes the LED's overall brightness drop significantly.

The epoxy compound comes in either liquid or solid pallets. Today, there are many players in the liquid epoxy compound supplier, for example, Dow Corning, Hankel, Delo, and many more. There are only a few solid pallets. The biggest supplier is Nitto Denko. Epoxy compound encapsulation gives excellent protection from corrosive environments. Its hardness is 84 (Shore D), and Young's modulus is 3200 N/mm^2 as illustrated in Figure 2.19 [18]. Its water absorption is just 0.13% when exposed to 24 hours of soaking at 25 °C [18]. Water is very detrimental to LED chips and can cause corrosion. Epoxy has good heat resistance. It has glass transition temperature (T_g) of roughly 125 °C, which is good for mid-power automotive applications.

Besides the "carbonization" effect, an epoxy compound is also not suitable for high-power LED applications because the heat generated from the LED chip will cause the epoxy encapsulant to crack due to stress in the epoxy. This is due to the CTE mismatch between the epoxy and chip. The CTE of epoxy is in the range of 60 ppm, and the chip is roughly 3 ppm. This big mismatch causes very high stress. The stress can be estimated using Eq. (2.5).

$$\sigma = E\alpha(T - T_{ref}) \tag{2.5}$$

where E = Young's modulus, α = coefficient of thermal expansion, T_{ref} = reference temperature and T = temperature.

Since stress is proportional to the CTE and Young's modulus, one can reduce stress in the epoxy-encapsulated package by either reducing Young's modulus or the CTE

Figure 2.18 LED epoxy package turned dark after prolonged exposure to blue light. Source: Courtesy of ams OSRAM GmbH.

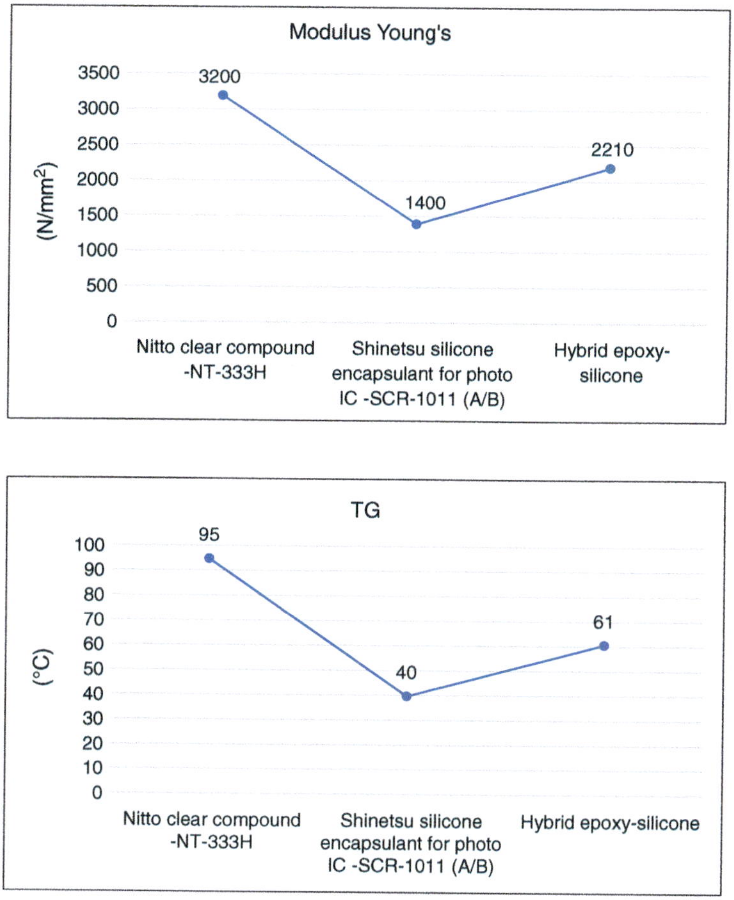

Figure 2.19 Mechanical properties of encapsulant materials. Source: Denko [18, 19]; ShinEtsu [20].

of the encapsulation material. There are ways to bring down these two properties; however, the material formulation has to be changed. One example to reduce the CTE was to use a filler with low CTE that had the same index of refraction as epoxy. This will bring the CTE lower. However, by doing this, the transmissivity may be reduced slightly. On the other hand, to reduce the Young's modulus, hybrid materials can be used, where epoxy and silicone materials are mixed to bring down the modulus. The Young's modulus of Nitto epoxy compound is usually around 3200 N/mm^2. This is quite high in comparison to silicone encapsulant material of 1400 N/mm^2, as shown in Figure 2.19. Hybrid, on the other hand, is at 2210 N/mm^2.

To compensate for the weakness in epoxy-based encapsulant material, silicone-based material was developed to support high temperatures and high energy light. This is important, especially for some mid- and high-power packages as the junction temperature can go beyond 125 °C. The disadvantage of silicone materials is that they have a too high CTE and low Young's modulus. As a result, they are soft and have low mechanical stability. They are sticky. This is an adverse effect in LED manufacturing, as this property causes the sticky silicone surface and tends to pick up particles and dust from the environment. Silicone material has high diffusion coefficients for gases, thus easily causing corrosion on the silver surface of the lead frame.

Silicone encapsulation gets a lot of attraction as its application in high-power and short-wavelength LEDs increases. The weakness of the epoxy compound allows silicone encapsulation material to fill the gap. Similar to the epoxy compound, silicone provides mechanical protection to the chip and wire. However, the Young's modulus of silicone is low, and the hardness is 70 (Shore D), lower than epoxy. It provides good heat resistance, is stable at high temperatures, and doesn't have any carbonization effect as the epoxy compound did because there is no carbon. Silicone encapsulation can be done either by casting or compression molding process.

2.2.2.2 Hermetic Sealed Package – Metal Can

Hermetically sealed packages, as shown in Figure 2.20, have been in the industry for more than 70 years, since the beginning of the LED industry. This package is hermetically sealed using the spot welding process in dry air or N_2-rich ambient conditions. Some metal-can packages have a lens, and some are not. It depends on the application. Metal can package is relatively expensive compared to standard PLCC or QFN LED packages as the materials used are glass, copper, and stainless steel. Hence, the application of metal can device is mostly for the harsh environments or laser product that require such a hermetically sealed package. Example: for automotive laser headlamp or gas sensor.

Beside metal can packages, there is also a butterfly package, as shown in Figure 2.21. This application is specifically for a high-power laser projector. The thermal load for this application is very high at 50 W. Hence, this "Butterfly" package is ideal for this application because it has a very good heat dissipation and good hermetic sealed that avoid the laser chip from degrading in the presence of atmospheric oxygen and moisture. By having good heat dissipation and hermetically sealed package, this package maintains very high reliability capability

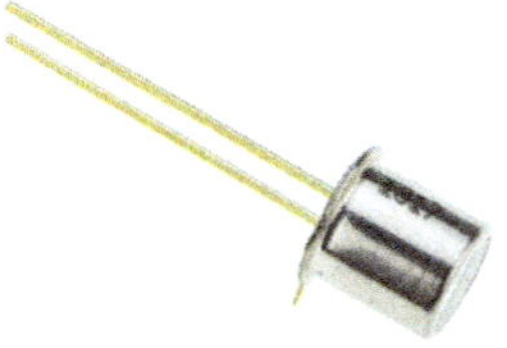

OSRAM Opto TO18, BPX 65

Cross-section view of metalcan device

Figure 2.20 Metal can package. Source: Courtesy of ams OSRAM GmbH.

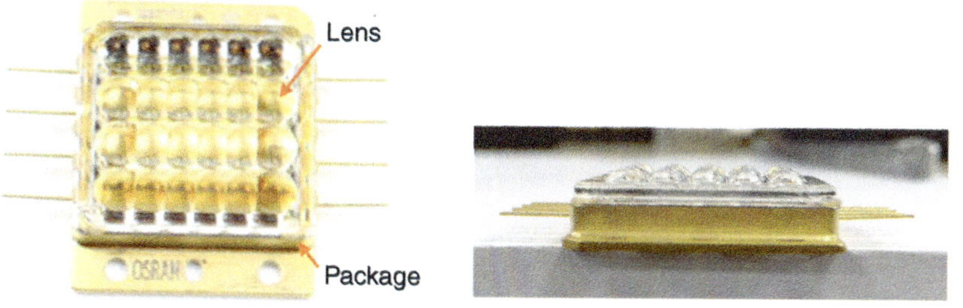

Figure 2.21 OSRAM Butterfly package for Laser project application. Source: Courtesy of ams OSRAM Group.

in harsh conditions like 85 °C and 85% relative humidity, with lifetime of more than 20,000 hours, making it ideal for the light source of a cinema projector. The sealing is done using spot welding process in an inert atmosphere. Usually in a rich nitrogen gas environment.

This package is relatively expensive as it is made of high-grade copper and thick gold-plated package. It provides very good heat dissipation and hermeticity.

2.2.2.3 Epoxy Cap Encapsulation

Epoxy cap encapsulation package, as illustrated in Figure 2.22, is a package that has an open-air package. The cap is mostly made of epoxy material. The cap design very much depends on the customer application. Some of the cap encapsulation is also directly attached to QFN package that has been bonded to the chip and wire, as illustrated in Figure 2.22a. There are also packages that are caped on PCB board that has bonded the chip and wire, as illustrated in Figure 2.22b. These types of packages are commonly used for mobile phone applications where the application condition is milder in comparison to automotive or industrial applications.

2.2.2.4 Glass Cap on Ceramic or Aluminum Encapsulation

Glass cap encapsulation is mostly used for very high-power LEDs, Laser diode, or UV LEDs. There are many players who supply the glass-cap encapsulated package.

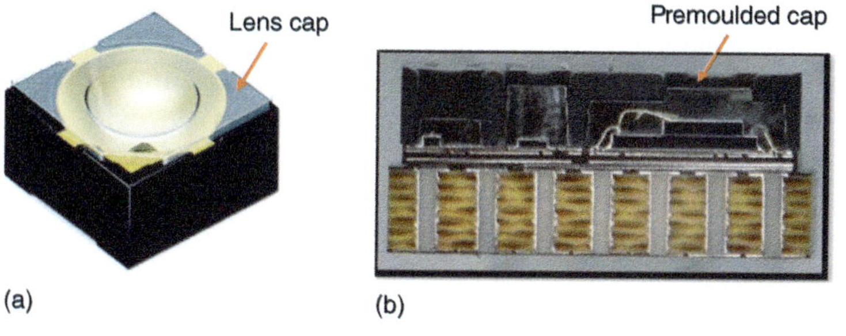

Figure 2.22 Epoxy cap encapsulation package. (a) Lens attached on the pre-moulded package. (b) SiP package, the cap is directly attached on PCB. Source: Courtesy of ams OSRAM GmbH.

Figure 2.23 ams OSRAM's new glass-encapsulated UV LED. Source: Courtesy of ams OSRAM GmbH.

One of them is OSRAM. OSRAM applied glass-sealed package technology for ultraviolet LEDs. This new product is completely sealed with the LED chip in the ceramic package and a glass cap to minimize the impact of gas or moisture on UV chip. This package, as illustrated in Figure 2.23, can maintain high reliability in many different environments – high temperatures up to 100 °C and high temperature and humidity environments (85 °C and 85% relative humidity). The glass cap is vital for UV-LED applications, especially for UV-C LEDs, where the radiation would degrade any carbon-based encapsulation. Hence, this package is ideal for UV-C LED packaging.

2.3 LED Thermal Management

We have generally explained the importance of thermal management to LED performance in earlier chapters. In Section 2.3, we will further explore in detail about the science of LED thermal management.

Good thermal management in LED is one of the key factors that guarantee the long life of an LED in its usage condition. An LED packaged with good heat dissipation design and low thermal resistance, will help to maintain the LED's long-term lifetime. Besides LED lifetime, the temperature increase in LED also causes brightness drop and the wavelength shift during the application. The LED wavelength will shift to a lower wavelength as the LED junction temperature increases. The color of LEDs also changes with the increase in temperature. In particular, phosphor-converted LEDs with blue InGaN and yellow phosphors experience light output degradation along with shifts of blue peak wavelength.

In LEDs, temperature increase results in forward voltage drop due to the decrease of the bandgap energy of the active region of LEDs and also results in the decrease in series resistance. The resistance decrease is due to higher acceptor activation occurring at the elevated temperatures as well as the resulting high conductivity at the active layers [1]. To understand further, we have to understand the fundamentals of LED thermal behaviors.

2.3.1 Fundamental of the LED Thermal Behaviors

The thermal behavior of an LED is undoubtedly one of the most important topics of research and main consideration for any LED application. It is well known in any LED industry that heat affects LED's performance and reliability.

It is common knowledge in LED industry that some of the energy is converted to heat. Hence, managing the heat in the LED is very crucial. One may ask how and where this heat is generated in the LED? How does the heat flow in the LED?

The main heat sources in the LED package are the heat generation at the junction due to the non-radioactive recombination processes, Joule heating at the series electrical resistance of the diode, and possible Joule heating at the interconnects. At the same time, the light absorption in the material and the interface is another heat source. The heat generated by LEDs has to be dissipated efficiently. One of the key parameters in understanding LED thermal behavior is to understand the temperature sensitive parameters (TSP). One of the TSPs is the forward voltage. When there is impedance at the thermal path in an LED, the forward voltage will be affected. In order to understand this phenomenon, one has to decipher the fundamentals of the LED's thermal behavior. To begin with, the fundamentals of the forwarding current and voltage of an LED have to be explored. Temperature increases in LEDs will affect the operation of the active layer and conductive layer in the LEDs. The forward current flows across the p–n junction when the anode-to-cathode voltage reaches a threshold value close to the diffusion voltage, V_D: $V_{th} \approx V_D$. This voltage value is usually published on the LED datasheets as "forward voltage." It is related to the bandgap energy of the LED material system [21]. The diffusion voltage should satisfy Eq. (2.6).

$$qV_D - E_g + (E_F - E_V) + (E_C - E_F) = 0 \tag{2.6}$$

where q, E_g, E_F, E_V, and E_C are elementary charge, bandgap energy, fermi, valence, and conduction band energy level, respectively, as shown in Figure 2.24.

In highly doped semiconductors, the separation between the conduction band edges (E_C) and fermi level (E_F) is small if compared to the bandgap energy $(E_C - E_F) \ll E_g$ on the n-type side and $(E_F - E_V) \gg E_g$ on the p-type side. Hence,

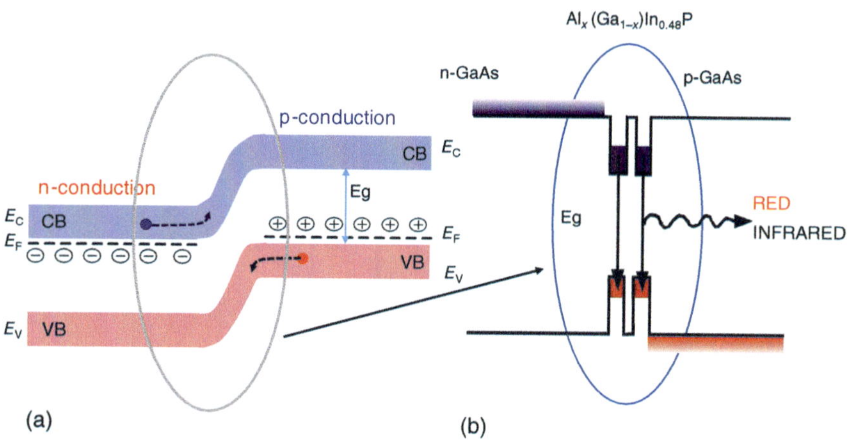

Figure 2.24 p–n junction under the forward bias conditions, minority carriers diffuse into the neutral regions, where they recombine (a) Simple LED with p-n-junction. (b) AlInGaP LED with quantum wells inside the p-n-junction.

by neglecting ($E_F - E_v$) and ($E_C - E_F$), the diffusion voltage can be approximated by the bandgap energy divided by the elementary charges, $V_{th} \approx V_D \approx E_g/q$ [1]. Using $V_D \approx E_g/q$, one can estimate the forward voltage of various LED wavelengths. GaInN (blue) has $E_g = 2.9$ eV, the forward voltage is 2.9 V. GaAs (red) has $E_g = 2.0$ eV, the forward voltage is 2.0 V. Thus, LED with a short wavelength emission has a higher forward voltage compared to one with a longer wavelength emission.

The forward current of LEDs is derived from Shockley's ideal diode characteristic as described in Eq. (2.7) [22].

$$I_F(V_F) = I_0\left(e^{(V_{Fp\text{-}n}/\mathbf{m}V_T)} - 1\right) \tag{2.7}$$

where I_0 is the saturation current of the ideal diode characteristic, $V_{Fp\text{-}n,}$ is the internal junction voltage of the diode, V_T is the thermal voltage, and **m** is a device-specific constant called the ideality factor. $I_0 = A{\cdot}q{\cdot}n_i^2\ (D_n/L_n{\cdot}N_A + D_p/L_p{\cdot}N_D)$. A is the cross-sectional area. D_n, D_p is the diffusion coefficient of the electron and hole. L_n, L_p is the diffusion length of the electron and hole. N_A, N_D is the dopant concertation of the free hole and electrons. q is the elementary charge. The thermal voltage V_T is described as $V_T = k{\cdot}T/q$. T is the junction temperature of the diode, and k is Boltzmann's constant [21].

The overall forward voltage V_F measured between the anode and cathode contacts of an LED is the sum of the voltage drop on the internal electrical series resistance V_R and internal junction voltage denoted by $V_{Fp\text{-}n}$, as shown in Figure 2.25. Hence, using the notations shown in Figure 2.25, $V_{Fp\text{-}n}$ can be described as in Eq. (2.8).

$$V_{Fp-n} = V_F - V_R = V_F - I_F{\cdot}R_s \tag{2.8}$$

by approximating $e^{(V_{Fp\text{-}n}/\mathbf{m}V_T)} - 1$, with $e^{(V_{Fp\text{-}n}/\mathbf{m}V_T)}$, Eq. (2.7) reads as follows:

$$I_F = I_0 e^{[(V_F - I_F R_S)/\mathbf{m}V_T]} \tag{2.9}$$

Adding I_0 and V_T, the forward current is read as:

$$I_F = A{\cdot}q{\cdot}n_i^2(D_n/L_n{\cdot}N_A + D_p/L_p{\cdot}N_D){\cdot}\left(e^{[(V_F - I_F\cdot R_S)/(\mathbf{m}kT/q)]}\right) \tag{2.10}$$

This equation shows the forward current dependence on the carrier concentrations.

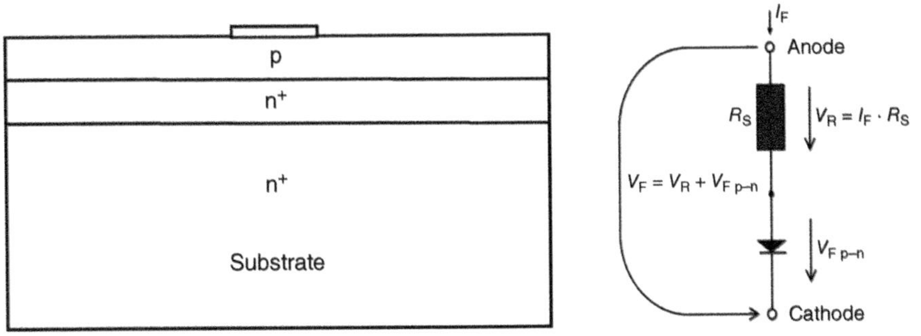

Figure 2.25 The basic LED construction with an equivalent circuit diagram of a nonideal diode with the internal p–n junction and the internal series electrical resistance. Source: Adapted from Lasance et al. [21].

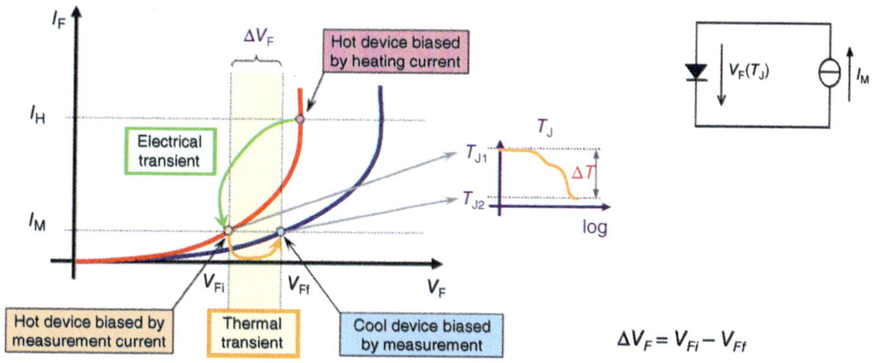

Figure 2.26 The electrical and thermal transient processes occurred during the testing of LEDs at a constant current [23].

Equation (2.10) also indicates that the LED parameters are strongly temperature-dependent. One of the most important factors determining the temperature dependence is n_i^2, which is the square of the intrinsic carrier concentration.

Figure 2.26 illustrates the temperature dependence of the forward voltage of p–n junctions when a fixed small forward current is forced across the junction. This electrical configuration is the most common in case of DC-driven LEDs. This current is denoted by I_M as the measurement current. With this configuration, the question of interest is the temperature dependence of V_F forward voltage.

$$\Delta V_F = V_{Fi} - V_{Ff}$$

The temperature sensitivity of the forward voltage of p–n junctions can be derived from the ideal diode characteristic expressed in Eq. (2.9). However, this has to comply with one condition, that is, the operation is in the range of small forward currents when the effects of the internal series resistance are still negligible [21]. Assuming that $(e^{(V_F/\mathbf{m}V_T)}) \gg 1$ and rearranging it in Eq. (2.9) for the forward voltage, one obtains the following:

$$I_F = I_0 e^{[(V_F - I_F R_S)/\mathbf{m}V_T]}$$

$$V_F = \mathbf{m}V_T \ln(I_F/I_0) + I_F \cdot R_s$$

when R_s is negligible $\approx 0\,\Omega$, the temperature-sensitive parameter that is V_F is also written as:

$$V_F = \mathbf{m}V_T \cdot \ln(I_F/I_0) \tag{2.11}$$

The temperature-dependent forward voltage is further derived using the differential method to obtain the sensitivity factor, S_{VF}, which is dV_F/dT_J. The final result of the derivation is as follows:

$$d\,V_F/d\,T_J = V_F/T_J - (V_{G0} + \mathbf{m} \cdot l \cdot V_T)/T_J \tag{2.12}$$

where T_J is the junction temperature of the LED, l is the power factor in the temperature dependence of n_i^2 in the formula of the I_0 saturation current. Typically, l value is around 3. V_{G0} is the nominal value of the bandgap voltage of the semiconductor materials, $V_{G0} = E_g/q$. The bandgap energy of the materials used in the LED is around 1.4–4.0 eV, the corresponding bandgap voltage is around 1.4–4.0 V. Assuming an ideal diode in the normal operation mode, $m = 1$, it results in Eq. (2.13) [22]:

$$S_{VF} = \mathrm{d}\,V_F/\mathrm{d}\,T_J = (V_F - 3\,V_T - V_{G0})/T_J \tag{2.13}$$

For single p-n-junction LED, this S_{VF} value is in the range of −1 to −3 mV/°C. The S_{VF} is also called temperature coefficient, α. The reciprocal of α is called K-factor, defined in the semiconductor device thermal testing standard is used for determining the thermal resistance of the LED as shown in Eq. (2.14) [23].

$$R_{th} = \mathrm{d}\,V_F/S_{VF}{}^{*}\,\mathrm{d}\,P_H \tag{2.14}$$

where

R_{th} = thermal resistance
$\mathrm{d}P_H = (P_H - P_M)$
P_H = heating power at hot condition ($I_H * V_H$)
P_M = heating power at measuring condition ($I_M * V_{Fi}$)

To determine an LED thermal resistance, T3Ster equipment is commonly used. Andras Pope et al., in their work on thermal resistance measurement, have elaborated in great detail on this matter.

2.3.2 Thermal Design in LED Package

Thermal design in LED packages has been one of the key components of success in LED's outstanding performance. A poor thermal design will make the junction temperature of LED stay at a high level, which may result in a low lifetime and bad reliability of an LED. Therefore, it is necessary to analyze the thermal characteristics of the LED and optimize its thermal design. In Figure 2.27, we illustrate a conceptual diagram of an LED thermal performance of OSRAM Golden Dragon® package. With the LED efficiency of 20–40%, roughly 60–80% of electrical power will be converted to heat energy. This heat has to be managed efficiently.

The input power is in the form of electrical power P_{el} is converted to optical and thermal power. This energy conversion can be equated as per Eq. (2.15). P_{th} is the thermal power, and η is the efficiency of the chip.

$$P_{th} = (1 - \eta) * P_{el} \tag{2.15}$$

The higher the chip efficiency, the lesser the heat generated. However, one has to understand that heat generated in the LED chip has to be dissipated efficiently to avoid LED overheating and failure associated with heat. To define whether an LED has good heat dissipation or not, the thermal resistance of an LED must be measured using the T3Ster equipment. Figure 2.28 gives a general explanation of the heat path in an LED and the thermal resistance and thermal capacitance.

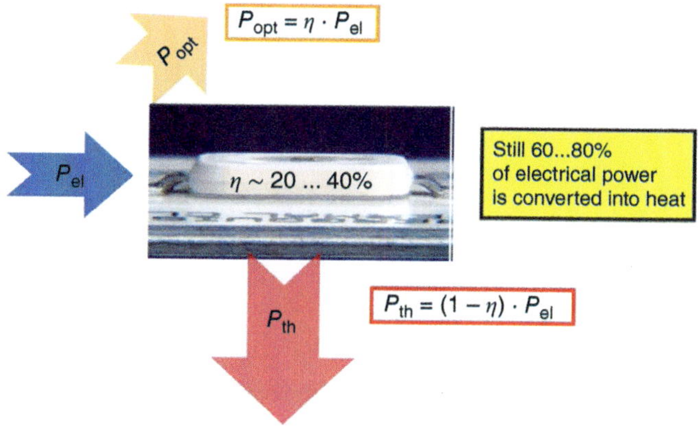

Figure 2.27 Shows OSRAM Golden Dragon package's thermal performance.

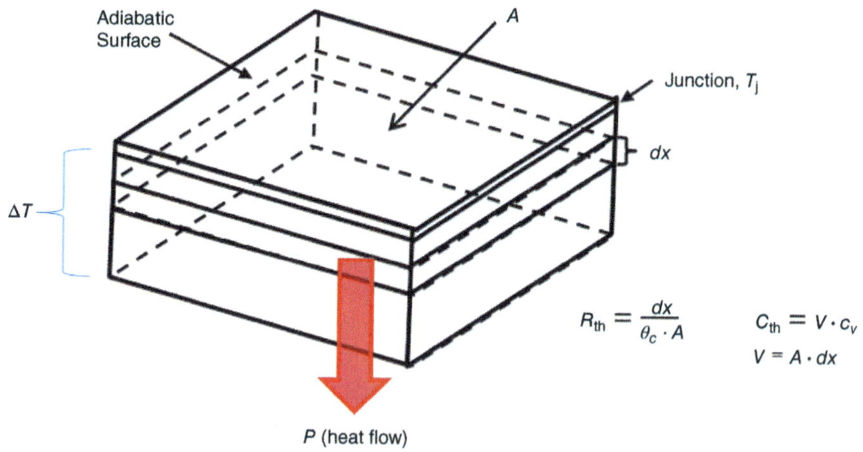

Figure 2.28 Conduction heat path in the LED die. It passes through the chip substrate and cracks.

The heat generated in the p–n junction will go primarily through the thermal path shown in Figure 2.28. There will be a temperature drop between the two isothermal surfaces of the material (assuming an adiabatic condition at the other surface) at ΔT. The R_{th} between the junction to the chip substrate can be described in Eq. (2.16), as explained below:

$$R_{th} = dx/\theta_c \cdot A \tag{2.16}$$

where θ_c is thermal conductivity, dx is the thickness, and A is the area size.

On the other hand, the chip's ability to keep the heat, which is the thermal capacitance, C_{th} between chip junction to die substrate, can be estimated using Eq. (2.17).

$$C_{th} = c_v \cdot V \tag{2.17}$$

c_v is the volumetric specific heat of die substrate (Ge), and chip volume, $V = A \cdot dx$.

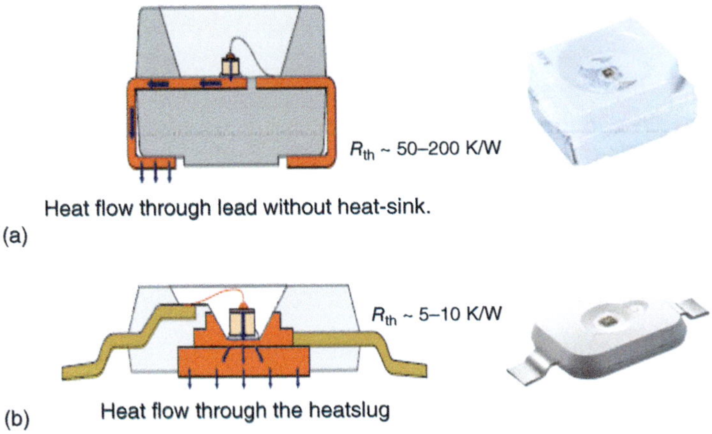

Figure 2.29 A comparison of thermal path and thermal resistance between low-power LED and high-power LED packages. (a) Standard surface mount LED package (OSRAM top LED® package). (b) High power LED package (OSRAM–golden Dragon® package). Source: Courtesy of ams OSRAM GmbH.

Since the chip is small, the thermal capacitance value of the chip is also relatively small in this case; hence, chip cannot keep the heat. As a result, the heat will directly flow through the die to package. Here, the package heat sink plays a very crucial role in dissipating heat. To solve this heat dissipation problem, the LED package has to have a good heat sink. Especially for high-power LEDs. Figure 2.29 shows a comparison of low-power LED and high-power LED package thermal management designs. Figure 2.29a shows a lower-power package from OSRAM, SMT TopLED, without heatsink, which has a thermal resistance of 50–200 K/W. On the other hand, Figure 2.29b shows a high-power LED package from OSRAM, Golden Dragon, which has an R_{th} of 5–10 K/W with a heatsink. Golden Dragon has heatsink directly below the chip, and the heat was directly dissipated in the shortest path to the external body. However, SMT TopLED has no heatsink. It's the only path of heat dissipation through the cathode lead, where the path is longer and narrower. It is not so efficient heat dissipation. Hence, this package cannot be driven with high current. It will harm its electro-optical performance and lifetime.

2.3.3 Impact of Thermal Behavior of an LED on Its Performance

One of the challenges in LED performance is LED self-heating. This is mainly occurring for high-power LEDs, where an increase in power density and the current level of LED chips. Heat is generated at junction due to nonradiative recombination processes and Joule heating. Temperature increases will affect the operation of the active layer in the LEDs. If the temperature increases to a high level, these layers may either degrade temporarily or even fail permanently. It is crucial to minimize the temperature increase with the proper design of LEDs, both at the chip level and package level, not to mention at the system level. Hence, the thermal management solution has to be correctly incorporated.

Self-heating also affects LED efficiency and efficacy. It has been shown that the thermal stress induced by self-heating in high-power LEDs can cause a decline in LED brightness and is generally considered to be one important factor in long-term reliability. The temperature at the junction can be estimated using Eq. (2.18). The junction temperature T_J is equal to temperature rise added to the ambient temperature, T_a. Temperature rise, on the other hand, is the multiplication of thermal resistance and heat power generated by the LED, $R_{th} * P_{th}$. The equation to estimate the heat power generation can be found in Eq. (2.15).

$$T_J = T_a + R_{th}{}^* P_{th}$$

where $P_{th} = (P_{el} - P_{op})$ and $P_{el} = I_f * V_f$.

$$T_J = T_a + R_{th}{}^* \left(I_f{}^* V_f - P_{opt}\right) \tag{2.18}$$

With Eq. (2.18), one can exactly estimate the junction temperature of an LED. Having known the junction temperature, one can carefully design the package to meet the desired junction temperature by modifying the thermal resistance of the package.

2.4 Electrical Contact Design

The electrical contact design in packaging technology is very critical to reduce energy loss and to avoid any heat generation due to ohmic resistance at the contact. The electrical contact of an LED package is divided into two categories. One is the internal electrical contact, which is where the chip is attached to cathode lead at substrate and wire bonding to that connects the chip and lead at the substrate as illustrated in Figure 2.30. This forms the internal electrical circuit in the package. The second category is external contact.

The external contact of an LED package is the external lead that will be soldered onto the external PCB. The lead design has to follow the application circuit board

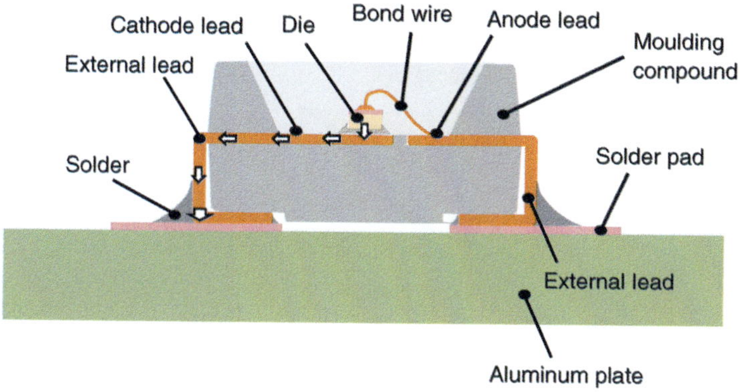

Figure 2.30 The internal and external electrical contact in the LED package. Source: Courtesy of ams OSRAM GmbH.

design. This external contact provides the electrical paths for power and signal distribution.

2.5 LED Light Conversion Principle

The light conversion technology is a uniquely innovative technology to convert short-wavelength emissions from the primary LED chip source to long-wavelength emissions by using phosphor conversion material. The phosphor layer plays an important role in determining the final optical performance of LED devices. Some of the studies have shown the effect of phosphor properties on packaging performance. The location, thickness, concentration, and geometry of the phosphor layer are the important factors affecting an LED's optical performance.

In one research work carried out by Kim et al., it was concluded that the reflector cup surface roughness, geometric dimension, and phosphor placement strongly influence the phosphor efficiency in the white LEDs. They found improvement in LEE by using the remote phosphor arrangement and a defused reflector, as shown in Figure 2.31a. This improvement is mainly attributed to a reduced re-absorption probability of wavelength-converted light by the LED chip and surrounding surface [24].

In another similar area of study by Yu et al. on the effect of the phosphor geometry on the luminous flux of LEDs shows that the remote phosphor configuration has higher luminous flux than the conventional dispersed-coating LED [25]. The remote-phosphor LED with a phosphor layer on a hemispherical top surface with convex bottom, as illustrated in Figure 2.31b can get a more than 12% improvement compared with conventional dispersed coating LED and a 5% improvement compared with the two-flat LED as shown in Figure 2.31a.

2.6 Summary

LED packaging is important not only for protecting the chip and the interconnection, i.e. wire and solder, but also because it enhances the light outcoupling from the

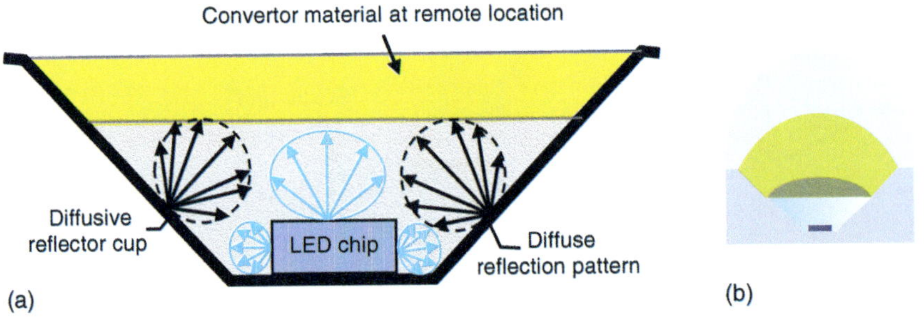

Figure 2.31 (a) Remote phosphor distribution of two-flat phosphor geometry in diffuse reflector cup, (b) remote phosphor distribution of hemispherical top surface with bottom convex phosphor geometry [24, 25].

chip. Packaging also helps to dissipate heat from the chip, which helps to improve the LED's lifetime and maintain brightness and color. Besides that, LED packaging also shapes the light output per the intended application.

References

1 Schubert, F.E. (2006). *Light Emitting Diode*, vol. 2. New York, USA: Cambridge University Press.

2 Taki, T. and Strassburg, M. (2020). Visible LEDs: more than efficient light. *ECS Journal of Solid State Science and Technology* 9 (1): 015017.

3 Zukauskas, A., Shur, M.S., and Gaska, R. (2002). *Introduction to Solid-State Lighting*. New York: Wiley.

4 Nakamura, S., Pearton, S., and Fasol, G. (2013). *The Blue Laser Diode: The Complete Story*. New York: Springer Science & Business Media.

5 Holonyak, N. Jr., and Bevacqua, S. (1962). Coherent (visible) light emission from $Ga(As_{1-x}P_x)$ junctions. *Applied Physics Letters* 1 (4): 82–83.

6 Haitz, R. (2003). Another semiconductor revolution: this time it's lighting! In: *Advances in Solid State Physics* (ed. R. Haitz), 35–50. Springer.

7 OSRAM (2016). LED fundamentals. http://ledlight.osram-os.com/knowledge/led-fundamentals/leds-basics (accessed April 2020).

8 Ganguly, A.K. (2007). *Optoelectronic Devices and Circuits – Theory and Applications*, vol. 2, 350. Oxford: Alpha Science International Ltd.

9 Nakamura, S., Pearton, S., and Fasol, G. (2013). *The Blue Laser Diode: The Complete StorySpringer-Verlag Berlin Heidelberg*. Berlin, Germany.

10 Krames, M. (2013). LIGHT-EMITTING DIODES: GaN-on-GaN platform removes cost/performance tradeoffs in LED lighting. *Laser Focus World* 49 (9): 37.

11 Vadim, S., Iktay, Y., Franko, K., and Hartnagel, H.L. (2017). Efficiency enhancement of InGaN/GaN LEDs with Mg-Si co-doped GaN quantum barrier. In: *Problems and Challenges of the Region's Economy in the Condition of Globalization*, 3e, 222–228. Comrat, Moldova: Instrumental Bibliometric National.

12 Fuji, T., Gao, Y., Sharma, R. et al. (2004). Increase in the extraction efficiency of GaN-based light-emitting diodes via surface roughening. *Applied Physics Letters* 84 (6): 855–857.

13 Wang, X., Lai, P., and Choi, H. (2010). The contribution of sidewall light extraction to efficiencies of polygonal light-emitting diodes shaped with laser micromachining. *Journal of Applied Physics* 108 (2): 023110.

14 Kashiwao, T., Hiura, M., Lim, Y.Y. et al. (2016). Optimization of surface-mount-device light-emitting diode packaging: investigation of effects of component optical properties on light extraction efficiency. *Optical Engineering* 55 (2): 025101.

15 Yang, S., Kwak, S.Y., Jin, J. et al. (2012). Thermally resistant UV-curable epoxy–siloxane hybrid materials for light emitting diode (LED) encapsulation. *Journal of Materials Chemistry* 22 (18): 8874–8880.

16 Pan, Y., Zhu, F., Fan, J. et al. (2018). Investigation of mechanical properties of silicone/phosphor composite used in light emitting diodes package. *Polymers* 10 (2): 195.

17 Huang, C.-H., Kang, C.Y., Chang, S.H. et al. (2019). Ultra-high light extraction efficiency and ultra thin mini LED solution by freeform surface chip scale package array. *Crystals* 9 (4): 202.

18 Denko, N. (2006). *Clear Epoxy Moulding Compound*. Osaka: Nitto Denko Corporation.

19 Denko, N. (2001). *NT300 Series Clear Epoxy Moulding Compound Handling Manual*. Osaka: Nitto Denko Corporation.

20 ShinEtsu (2005). *ShinEtsu SCR-1011(A/B) Silicone Encapsulant*. Tokyo: ShinEtsu Corporation.

21 Lasance, C.J.M. and Poppe, A. (2014). *Thermal Management for LED Applications*, 541. New York: Springer.

22 J.S.S.T. Association. (1995). Intergrated circuit thermal measurement method – electrical test method (single semiconductor device). *JEDEC Standard JESD51-1*. © JEDEC Solid State Technology Association, 2011.

23 J.S.S.T. Association. (2012). *JESD51-51*. Implementation of the Electrical Test Method for the Measurement of Real Thermal Resistance and Impedance of Light-Emitting Diodes with Exposed Cooling. JEDEC Solid State Technology Association: Arlington, VA. https://www.jedec.org.

24 Kim, J.K., Luo, H., Schubert, E.F. et al. (2005). Strongly enhanced phosphor efficiency in GaInN white light-emitting diodes using remote phosphor configuration and diffuse reflector cup. *Japanese Journal of Applied Physics* 44 (5L): L649.

25 Yu, R., Jin, S., Cen, S., and Liang, P. (2010). Effect of the phosphor geometry on the luminous flux of phosphor-converted light-emitting diodes. *IEEE Photonics Technology Letters* 22 (23): 1765–1767.

3

LED Packaging Manufacturing Technology

LED packaging technology is central to the LED manufacturing industry. For stable and consistent quality, an in-depth knowledge of LED package manufacturing technology is essential. Much progress has been achieved in improving the process efficiency, device efficiency, and material efficiency of LED packaging by the industry players; however, not much is available in the public domain of science and engineering journal articles as they belong to the intellectual property of the LED packaging manufacturers. Given this situation, writing this book on LED package manufacturing technology requires a great deal of maneuvering and, in certain cases, requires approval from relevant parties. In the next few subchapters, the LED package manufacturing technology will be detailed.

3.1 LED Packaging Process Flow

One of the elements of LED packaging technology is the manufacturing process. The LED packaging processes depend on the package design and intended uses. There are many types of LED packages and their intended uses. Hence, in this book, we only cover generic LED packaging processes. The generic LED packaging process flow is shown in Figure 3.1. This generic LED packaging process shows that the process starts with the die-attach process and continues with glue curing. Upon glue curing, a wire-bonding process takes place. The next process is encapsulation and then post-mold curing. Once the post-mold curing process is complete, the singulation process takes place before the final test and automated visual inspection process. The final process in the LED package manufacturing process is the packing process. Here, the LEDs are usually packed in moisture barrier bags before shipping to the customer.

3.1.1 Die-Attach Process

Generally, all LED packaging processes begin with a die-attach process. Here is where the die is attached to the substrate. The substrate can be a copper-lead frame, printed circuit board (PCB) or ceramic printed substrate, as shown in Figure 3.2a. The substrate selection depends on the package design and the end

LED Packaging Technologies: Design, Manufacture, and Applications, First Edition.
Luruthudass Annaniah, Mohamed Salleh M. Saheed, and Rajan Jose.

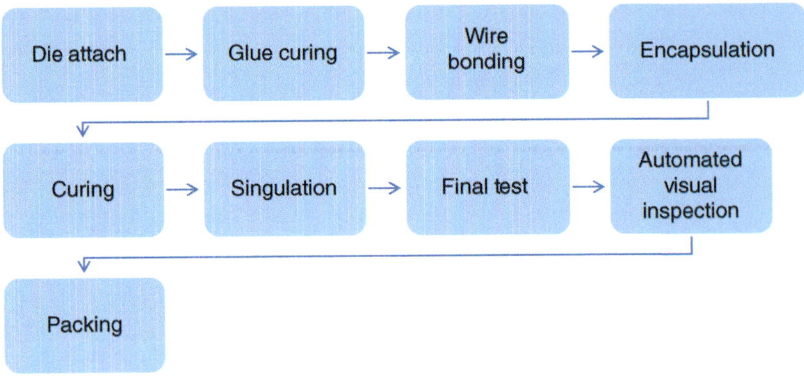

Figure 3.1 LED packaging process flow.

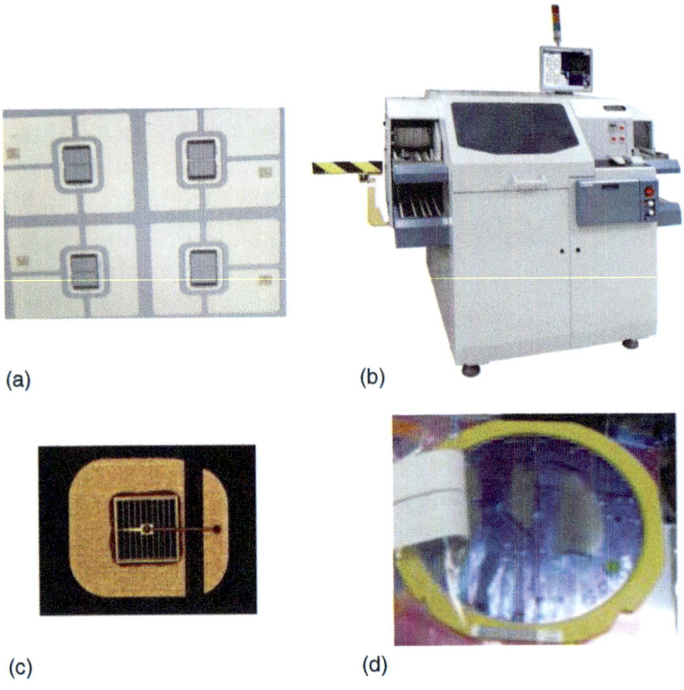

Figure 3.2 Die-attach process: (a) ceramic printed substrate, (b) die-attach equipment, (c) die-attached on the substrate, (d) dies on mylar. Source: Courtesy of ams OSRAM GmbH.

application requirements. There are mainly two types of die-attach processes: (i) die-attach process using glue attach and (ii) die-attach using the soldering process. Usually, low- and mid-power LED packages use a glue die-attach process. On the other hand, most high-power LED packages use the die-attach soldering process because of their high reliability. This is mainly due to good intermetallic interconnection that directly improved the thermal conductivity compared to the die glue interconnection. The glue-attached die's (chip) thermal conductivity is not as good as that of the soldered chip. However, for simplicity in describing the LED manufacturing process, we use the glue die-attach process.

3.1.1.1 Die-Attach and Glue Curing Process

Almost all the die-attach processes in the LED manufacturing industry use automated die-attach machines. Many industrial-grade die-attach machines vary in price and die-bonding tolerance accuracy. On an average, the die bonder's bonding accuracy with tolerance is ±30 μm. Usually, the unit per hour (UPH) output of LEDs can range from 5000 to 10,000 hours. Some cases even reach 20,000 UPH. Any further need for higher bonding accuracy, meaning tolerance < ± 30 μm, for example, ± 10 μm, the equipment price drastically increases, and the speed drastically slowed. Hence, the UPH output will drop drastically. In this compromised situation, only high-accuracy die placement applications will use such high-accuracy equipment. Today, most die-attach equipment uses average die-bonder accuracy for their mass production. The glue die-attach bonding equipment, shown in Figure 3.2b, is an industrial-grade bonder. In the die-attach process, the glue is dispensed onto the substrate before the die is picked up from mylar and placed on the substrate, as shown in Figure 3.2c.

The mylar, as shown in Figure 3.2d, usually has many dies, sometimes several thousand dies. In most cases, these dies were sorted for their brightness, voltage, or wavelength before placing them onto the mylar in the wafer manufacturing process.

The glue can be conductive for a die that has a conductive backside. For a nonconductive die backside, like a sapphire chip (die), where the anode and cathode contacts are on the surface, the glue in use is transparent nonconductive glue. On the other hand, the standard non-sapphire chip (die) uses silver or gold-filled glue. The glue coverage for the conductive glue must be lower than the p–n junction (active layer), as shown in Figure 3.3, to avoid shorts. Usually, the rule of thumb is that the glue should be controlled to roughly half of the die height. At the same time, the glue bond-line thickness is critical for the die attach strength. Having too little glue will compromise the mechanical strength of the die attached to the substrate. Bond-line

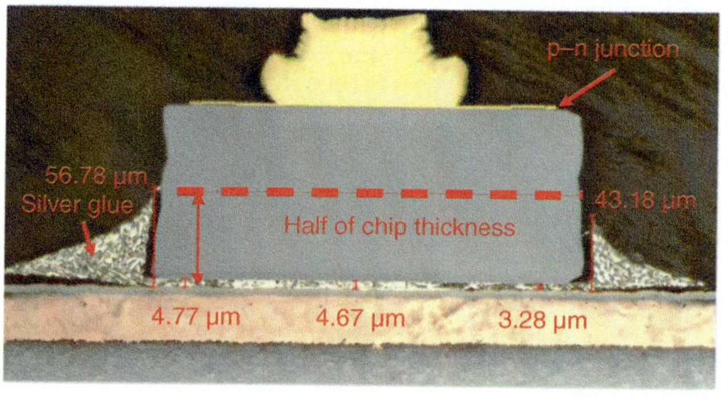

Figure 3.3 Small LED chip bondline thickness and glue (silver) height. Source: Courtesy of ams OSRAM GmbH.

thickness varies for different die sizes. Usually controlled in the range between 3 and 7 µm. Figure 3.3 shows a small LED die bond-line thickness and glue height, which were controlled at a range of 3.28–4.77 µm, and the glue height is at almost half die height, that is, a range of 43.18–56.78 µm.

After the die-attach is completed, the glue must be cured to harden and make the die permanently attached to the substrate. The glue curing process is done in a curing oven at a specified temperature, and the curing time is recommended by the glue supplier. Most of the glue curing process is done in a "Cold in – Cold Out." "Cold In – Cold Out" mode means that the die attached to the substrate is put into the oven at room temperature, and then the oven will be heated up. After it completed the curing process, the oven automatically will switch off, and it will cool down to room temperature before the substrate (with die attached) is taken out. There are also "Hot in – Hot out" glue curing processes practiced in the LED manufacturing industry. "Hot in – Hot out" means the die-attached substrates were put in a curing oven at curing temperature and taken out at the same curing temperature.

3.1.2 Wire Bonding Process

Wire bonding is one of the main processes for creating electrical interconnection between die and substrate. There are many wire bonding machines in use for the wire bonding process. Examples are Shinkawa, ASM, K&S, and ESEC. They have their advantages and disadvantages. The LED manufacturer chooses based on their needs. Off lately, equipment price has begun to be one of the key factors in equipment selection. This is due to stiff price erosion in the LED industry, which affects the margins of the LED manufacturer.

The wire used for LED wire bonding is mostly gold wire and, in some cases, aluminum wire, silver wire, or even copper wire, where product price is the main factor for business success. Gold wire is more expensive compared to aluminum or copper wire. However, gold is corrosive-resistant and has very low electrical resistance compared to aluminum wire. Figure 3.4 shows the fusing current comparison between gold, aluminum, and copper wire. Fusing current is defined as the minimum current at which the wire melts and disconnects the circuit. The wire size is dependent on the current density that is required to flow through the wire. This is related to the resistance in the wire. The higher the resistance in the wire, the lower the fusing current is. The fusing current is simplified by W.H. Preece in an equation shown in Eq. (3.1):

$$I_{\text{fusing}} = \theta d^{3/2} \tag{3.1}$$

where θ is a constant, depending on the material the wire is made of, and d is the diameter of the wire.

The wire selection is critical to the quality and reliability of the LED package. Besides that, LED applications also play an important role in wire type selection. For the automotive industry, the conditions the LEDs were exposed to are much harsher than in the consumer industry. Hence, gold wire is preferred, as it is more durable and resistant to corrosion. On the other hand, for the consumer industry, where the

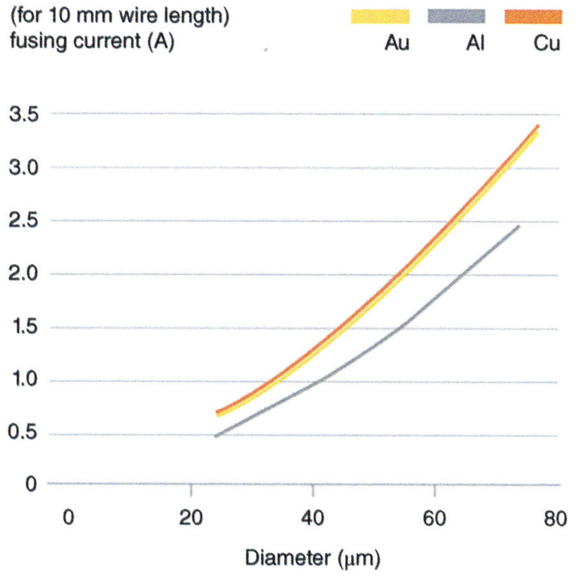

Figure 3.4 Gold wire against fussing current.

Figure 3.5 Gold wire bonding on sapphire chip. Source: Courtesy of ams OSRAM GmbH.

price-sensitive market is especially indoor lighting, aluminum wire is preferred by some LED manufacturers as it is more cost-effective.

There are many types of wire bonding loops. Figure 3.5 shows two types of wire bonding loops. The most common wire bonding loop is forward bonding. This wire bonding loop is used in most packages where the space for such bonding is insufficient. The second bonding shows the reverse bond ball stitch on the ball (BSOB) bonding. This bonding is usually used for small and thin packages where the space is limited.

One of the key factors that influence the wire-bonding quality is the heat-affected zone (HAZ) and the wire looping above the neck region, as shown in Figure 3.6a. Grain size at HAZ is relatively larger than the grain size of normal wire, as shown in Figure 3.6a. Gold wires for ball bonding are supplied in the annealed condition. During wire bonding, the wire at the portion immediately above the ball would become further annealed during wire melting and ball formation. The HAZ would thus become much weaker than the rest of the wire. Bending sharply above the ball as shown in Figure 3.6b will allow easy break-off at the neck. Even with

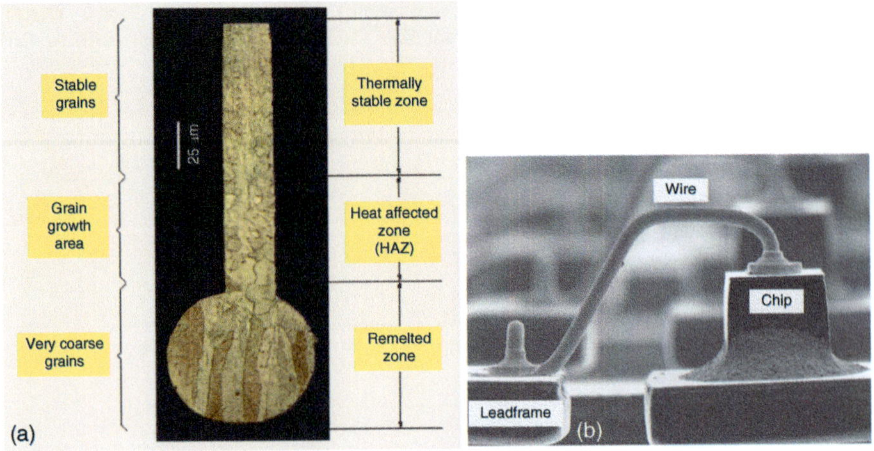

Figure 3.6 (a) Heat-affected zone (HAZ) of gold wire Source: Image from Heraeus Holding. (b) Wire looping. Source: Courtesy of ams OSRAM GmbH.

high-strength wire (with dopant in wire) this zone is still the weakest part of the wire bonding system. The hardness of this HAZ is roughly 20% lower than the rest of the wire. At wire pull test wire usually broken at HAZ area (neck). Many cases have been observed of the wire breaking at the HAZ area when subjected to extreme temperature cycles testing, for example, for automotive reliability testing at −40 and 125 °C for 2000 cycles.

Since the length of HAZ is extremely important, one has to control the HAZ formation in the wire bonding system. The HAZ is determined by both the rate of heat transfer to the wire tip and the heat flux along the axis of the wire. The most important parameters governing the melting and resolidification of the gold wire are the wire diameter (d), the discharge current (I), the gap (G) between the electrodes, and the duration of discharge (t). Hence, controlling these parameters will influence the HAZ.

3.1.3 Surveillance Checking Using Statistical Process Control

Surveillance checking is one of the keys to process control as a line of defense in the LED manufacturing process. One of the common surveillance checks is using statistical process control (SPC). It is crucial to check the die-attaching and wire-bonding performance of each manufacturing lot. Usually, it is done in a sampling of 5–10 pcs for each "Lot" or a certain defined interval of each process. A "Lot" refers to the number of items manufactured in a single production run. In the surveillance checking process, the LED die-attach strength and wire bonding strength are checked using Dage tester, as shown in Figure 3.7a. There are two types of tools used: one is a shear tool and the other is a wire-pull tool, as shown in Figures 3.7b,c, respectively.

The quality of die bonding is determined by the die shear strength. SPC samples were checked for die shear strength, and these data were computed in a control chart to see the values over different LED manufacturing lots, as shown in Figure 3.7d. If

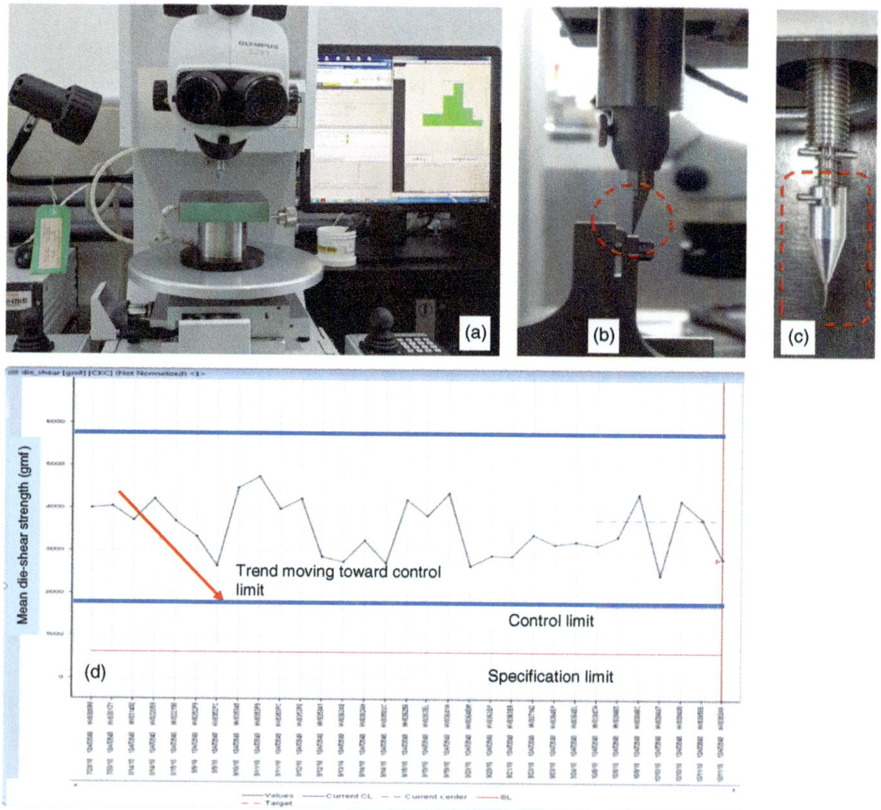

Figure 3.7 (a) Dage Tester, (b) the shear tool, (c) the wire pull tool, and (d) die-shear SPC diagram. Source: Courtesy of ams OSRAM GmbH.

the trend is continuously moving toward control limits, either upper or lower limits, it indicates something is not right (out of control). An indication of an out-of-control process is if five consecutive data points continuously move toward the upper or lower spec limits. In this situation, usually, the process engineers quickly investigate the die-attach process and the material used for the process to remedy this situation. Necessary containment action will be taken, and root cause findings will be initiated to remedy the situation.

Die shear strength is directly proportional to die size. The larger the die size the greater the die shear strength. Hence, in the LED manufacturing process, well-established control limits are used for different die sizes.

Besides the die shear strength, the glue remnant on the substrate is also checked. Usually, if there is some glue remaining on the substrate, as illustrated in Figure 3.8a, it indicates die to attach has good adhesion to the substrate. In case there is no glue remaining on the substrate, one must reconfirm back if there is any contaminant on the surface of the substrate, even though the shear strength value is within the control limits. This is a common practice in the LED packaging industry.

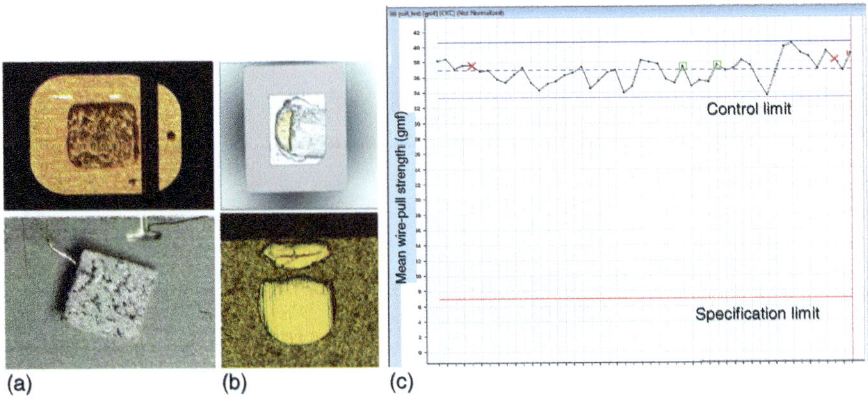

Figure 3.8 Surveillance checking, (a) glue remnant on the substrate. (b) gold remnant on die-pad and substrate surface, (c) Wire-pull SPC chart. Source:Courtesy of ams OSRAM GmbH.

The wire pull test is another surveillance check used in the LED packaging industry to ensure the wire is properly bonded to the die and substrate. Besides the pull test, a ball shear test is also conducted for the wire bonding process to check the ball shear strength. Gold remnant on the die bond pad, as shown in Figure 3.8b, is a crucial indicator of the intermetallic formation between gold wire and die bond pad. Having no gold remnant indicates poor intermetallic formation during the wire bonding process. Usually, it yields poor ball shear value.

Like the die shear test, the wire pull test is also monitored for its pull strength using a control chart, as shown in Figure 3.8c. This chart is closely monitored, and if the trend is moving toward the control limits, it indicates the wire bonding process is not optimum and there could be some issues.

3.1.4 Encapsulation Process and Post-Mold Curing Process

The encapsulation process is one of the most important processes in the LED packaging technology. Encapsulation by the name itself explains a form of mechanical protection. Besides that, encapsulation also defined the package outline and the light out-coupling efficiency and radiation characteristics. That means the radiation pattern of an LED can be designed into the package through the mold design. The light from the chip can be guided through the package to have maximum out-coupling efficiency.

There are many types of encapsulation processes. In this book, we will only cover the transfer molding process. In this molding process, transfer molding equipment and a transfer mold were used for the encapsulation process.

The transfer molding equipment shown in Figure 3.9a has a mold that consists of upper and lower mold chassis, as shown in Figure 3.9b. The transfer molding is shown in Figure 3.9c, which has plunger, transfer pot, mold cavities, runner, and ejector pins. The mold temperature recommended by the epoxy supplier is usually

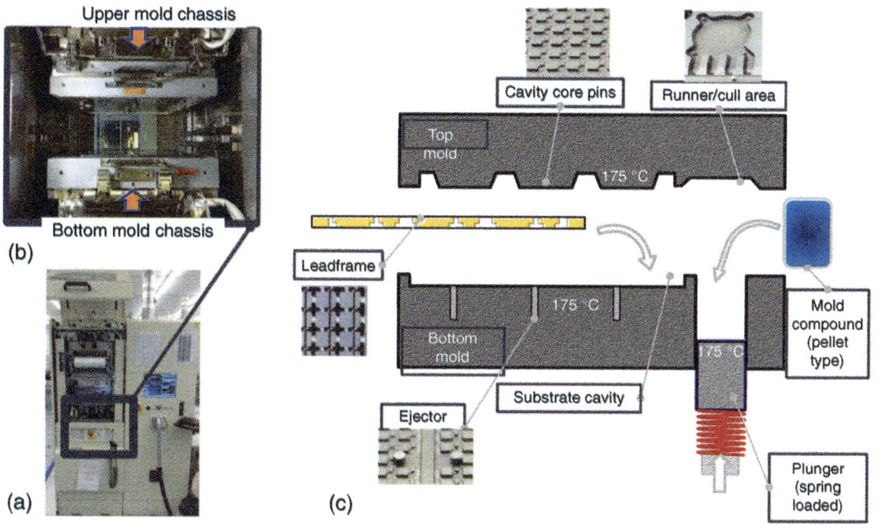

Figure 3.9 (a) Mold design with lens, (b) Transfer Molding equipment. (b) The upper and bottom part of mold. (c) Transfer molding concept showing general construction of mold. Source: Courtesy of ams OSRAM GmbH.

in the range of 150–160 °C for clear mold compounds. The mold compound is placed in the transfer pot. The epoxy mold compound softens at 100 °C and melts at roughly 150 °C. The plunger pushes the epoxy compound through the runners to the mold cavities. It takes roughly 120s to 150s for the epoxy compound to cross-link and harden in the mold cavities. Once this is passed, the mold will be opened, and the ejector pin would push the molded LED package out of the cavities. The LED packages are removed from the mold, and the mold cavities are cleaned before the next cycle of molding takes place.

After completion of the molding process, the molded LED packages must undergo a post mold curing process. This is to complete the epoxy cross-link process inside the package. It is commonly understood that during the molding process, roughly 85% of the epoxy is cross-linked. Hence, it is important to have a post-mold curing process. Usually, this is done at 150 °C for roughly five hours, as recommended by most epoxy compound suppliers. Failing to do this process will have severe mechanical defects in the LED package. Some of the common problems observed are cracked packages during LED application.

Since the LED requires transparent material as an encapsulant, the epoxy compound is usually pure epoxy without fillers. This enables light transmission above 90% [1]. Most of the industrial-grade epoxy compounds meet this requirement. This is important to minimize photons from the chip absorbed by the encapsulation material, which reduces package out-coupling efficiency and thereby reduces the LED brightness.

The mold cavities are carefully designed to meet the desired radiation pattern. Usually, optical simulation tools were used to design the mold tool cavities. Here, the encapsulant material's optical and mechanical properties were considered in designing the mold cavities. Epoxy compounds have an index of refraction of roughly 1.55, and silicone compounds have an index of refraction of roughly 1.50. This means silicone will have higher light refraction, which corresponds to a higher light loss in comparison to epoxy. Hence, the mold design for silicone encapsulant must consider this light loss.

3.1.5 Singulation Process

The singulation process is the process where the LED package is cut into a single LED from the substrate panel or leadframe. One of the common singulation processes used for ceramic or QFN panels is the sawing process. Here, basically, the workpiece (substrate panel), which is fixed on a mylar tape, as shown in Figure 3.10a, is cut using sawing blade. This panel on the mylar is fixed to the sawing table using a vacuum suction chuck. This fixing is to hold the substrate panel firmly during the sawing process. The sawing blade is sawing through the substrate, as shown in Figure 3.10b, until it partially penetrates the mylar. Hence, the mylar is still intact to hold the singulated LED units. The UPH or speed of sawing process varies with some cases going up to 10,000 UPH. In some cases, just 1000 UPH. This depends on the substrate material and sawing tolerance. Thick and brittle substrate material requires a slower speed. Sawing tolerance also influences speed. Usually, small sawing tolerance requires slower speed, as it requires a gentle sawing process. High-speed sawing usually causes chipping and affects final product's quality.

3.1.6 Final Test and Auto Vision System Process

The LED testing process is one of the key processes to ensure that LED parts supplied to the end customer are in accordance with the quality stipulated in the LED datasheet. Hence, the process must have the right equipment to sort the good from bad LEDs. Here, all the LEDs must be tested. The testing process is usually handled by a test handler that is integrated with an electro-optical tester and auto-visual inspection system. Figure 3.11a shows the industrial-grade test handler that is combined with an electro-optical tester and auto-vision system and the taping station as the output of the test handler.

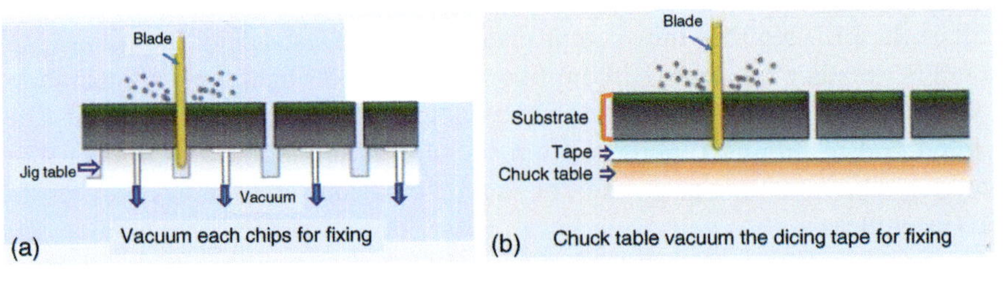

Figure 3.10 (a) Sawing process conceptual diagram. (b) Singulation blades grinding through the substrate. Source: Courtesy of ams OSRAM GmbH.

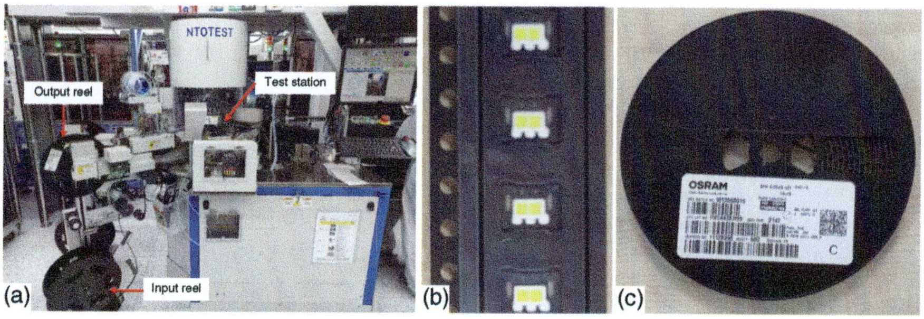

Figure 3.11 (a) An industrial-grade test handler with electro-optical tester and auto-vision system, (b) career tape with LED inside, (c) reel. Source: Courtesy of ams OSRAM GmbH.

Testing of LEDs follows a certain testing sequence, which usually starts first with the Kelvin contact test and is then followed by an electrical and optical test. The electro-optical test usually segregates the good parts into several good bins in accordance with testing specifications. These good bins have barcode labels that indicate their sorted electro-optical bins. These bins' specifications are stated in the product data sheet.

Once, these electro-optical tests are completed, the LEDs are visually checked by an auto-vision system for their mechanical and visual appearance. Upon completion of all these steps, the good parts will be taped onto a career tape, as shown in Figure 3.11b. The bad parts are binned in a reject bin for further failure analysis before they are scrapped.

The auto visual inspection (AVI) process is incorporated inside the test handler, which inspects the tested goods units. This inspection process not only checks the unit on the tape but also checks the taping quality. The inspected and sealed carrier tape, as shown in Figure 3.11b, is rolled in a reel. Once the reel is full, the equipment triggers an alarm for reel changing. The equipment operator removes the full reel and puts a new empty reel in the test equipment. The full reel, as shown in Figure 3.11c, then goes for the next step, which is the packing process.

3.1.7 Packing Process

The packing process must follow a strict industrial standard that is stipulated in JEDEC-STD-020C. It explains the moisture sensitivity requirement of the packing process. The first step in the dry packing process is to remove any moisture that has built up in the package. This is done by baking the finished product for 2.5–48 hours at temperatures between 85 and 125 °C, depending on the package type. During baking, the product is contained in high-temperature resistant device trays, aluminum trays, or tubes before testing these devices in the Final Test process. Within 24 hours after baking, the product must be sealed with a prescribed number of desiccant pouches and an indicator card in a dry bag under a partial vacuum [2].

There are several moisture levels, as shown in Table 3.1. The moisture sensitivity level (MSL) relates to the packaging and handling precautions for semiconductors. It applies to LEDs as well.

Unpacked devices may be mounted under environmental conditions not exceeding 30 °C and humidity levels of 60% RH. Devices must be soldered on a PCB assembly within specified floor life hours, as described in Table 3.1 [2].

Figure 3.12 describes the packing components in packaging of an LED product. Here, the labeled reel is packed into a moisture-sensitive barrier bag containing humidity indicator and desiccant. The moisture barrier bag also has a similar barcode label as that on the reel. All the moisture-sensitive barrier bags have a moisture-sensitive label on them to indicate the handling of these bags. The reel with the desiccant and moisture indicator is vacuum sealed and packed in a bigger corrugated box before shipping to the customer [2].

Table 3.1 Moisture sensitivity level and the requirement for packing.

Moisture sensitive level	Floor life
1	No limit
2	1 yr
2a	4 wk
3	168 h
4	72 h
5	48 h
5a	24 h
6	6 h

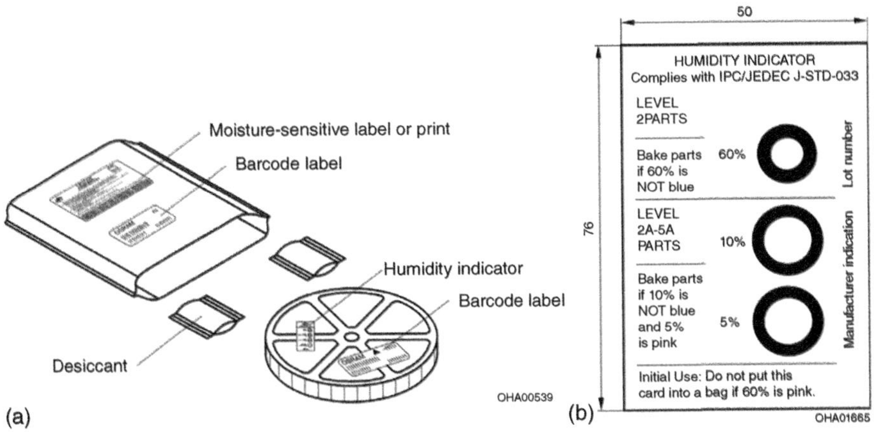

Figure 3.12 (a) Moisture-sensitive product is packed in a dry bag containing desiccant and a humidity card, (b) Humidity indicator card. Source: Courtesy of ams OSRAM GmbH.

3.2 Common Defects in LED Packaging Industry

There are many types of LED defects in the LED packaging industry. In this subchapter, some of the common defects and their impact on the electro-optical properties of LEDs are discussed.

3.2.1 Die-crack: Impact on the Electrical and Optical Properties of LED

Die-crack is one of the common failures observed in the LED packaging industry. There are many causes for die-cracks. One of the common causes is during the die-bonding process. A high bond force will lead to a die-crack. Sometimes high wire-bonding forces will also lead to die-crack, especially at the wire-bond pad. Besides this, die-crack can also be originated from the wafer sawing process.

The impact of die-crack on LED electro-optical properties is severe and usually detrimental to LED applications. Hence, this has to be compulsorily avoided.

Cracked die LED shows a distinct character when the LED is forward-biased using a curve tracer. The ideal characteristic of an LED can be described as shown in Figure 3.13. However, when the LED is cracked, it behaves differently, where it causes shunt, as shown in Figure 3.13a, or has a series of resistance, as shown in Figure 3.13b.

Another common defect in LEDs that causes shunt is the cracked die that affects the epitaxial layer, as shown in Figure 3.13a, presented by Ching et al. [3]. On the other hand, the series resistance effect is commonly caused by an LED interconnection defect. For example, the stitch bond pad is cracked and separated from the main die body, as reported by Lu G. et al. [4]. The LED failed soon after the stress test.

When the LED is biased in the reverse mode using a curve tracer, it shows a distinct characteristic, as shown in Figure 3.13c. In the LED industry, this reverse bias mode is also called as leakage current measurement.

Synonymous with the word "leakage current," when the reverse current of an LED is lower than the ordinary spec value, this indicates that there is a leakage in the LED that reveals the LED to be defective. In the LED testing environment, the reverse current test is a very common test parameter.

The main goal of LEDs is to produce light. However, the light output from the LED can be affected by a cracked die. As mentioned in the earlier section, when the epitaxial layer of LED cracks, the brightness drops, and in the worst situation, the LED will cease to operate.

The cracked die can directly affect the optical properties. A cracked die reduces the brightness and, at the same time, reduces the efficacy of the LED. This is observed in a real operation, as illustrated in Figure 3.14, which shows the dimming effects of the die crack in the LED.

Optical radiation pattern, on the other hand, is an important element for an LED user. Certain applications require a narrow viewing angle. Some may want otherwise, it depends on the end customer's needs [5].

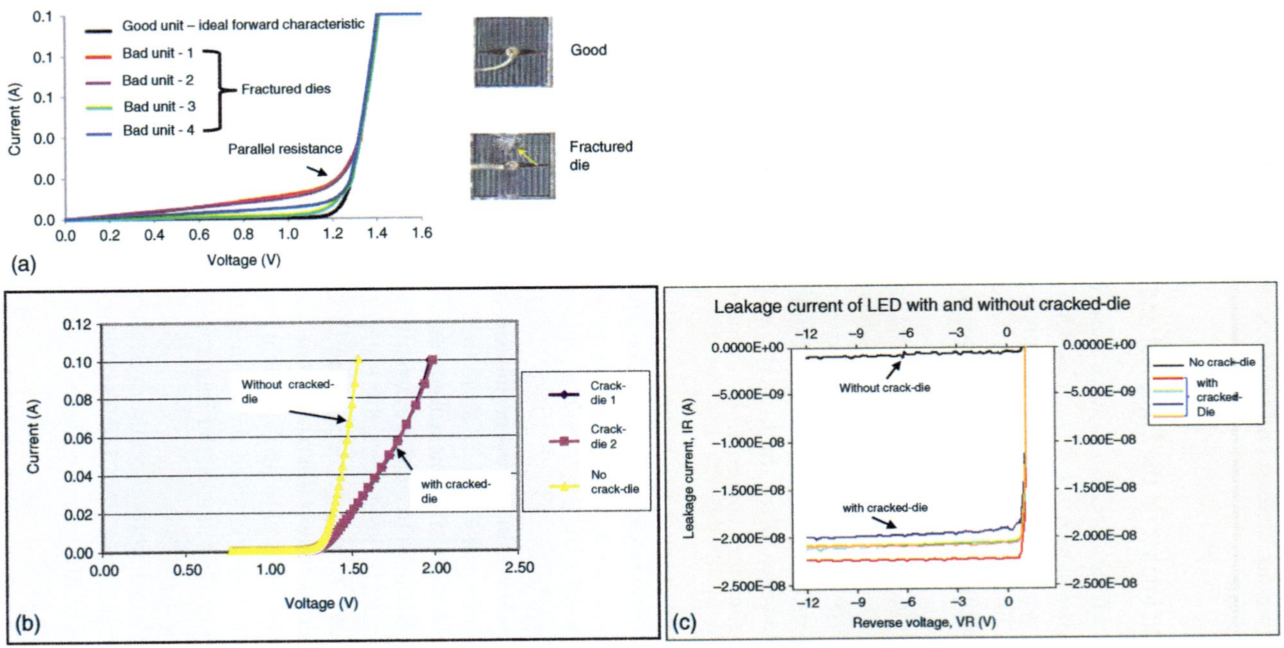

Figure 3.13 Current–voltage (*I*–*V*) curves for the selected fractured dice and good units. It shows a parallel resistance of the fractured dice. Source: Courtesy of ams OSRAM GmbH.

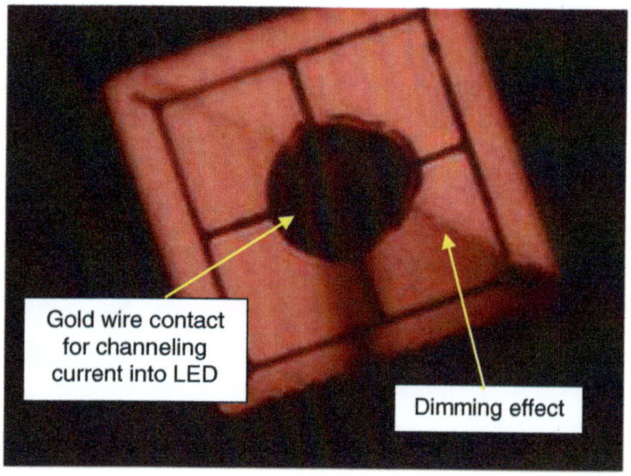

Figure 3.14 Die cracked LED biased with low current, showing a dimming affect along the crack line. Source: Courtesy of ams OSRAM GmbH.

To a certain extent, some crack events may affect the optical radiation pattern. The crack on the die can be so severe that the light path on the surface of the die is reflected at a certain angle that affects the optical radiation pattern. Hence, it is important to avoid cracks in the LED [6].

3.2.2 Lifted Die or Glue: Impact on LED Thermal Behavior and LED Performance

Some of the common LED failures incurs in LED manufacturing are lifted die or lifted glue failure, as shown in Figure 3.15a, b, respectively. They are mainly due to stress in the package or surface contamination. If the substrate, for example, the lead frame surface, is contaminated, it is very likely that a lifted glue issue will occur. On the other hand, if the die backside is contaminated, then the chances of the lifted die will be more likely to happen.

It is very difficult to detect the lifted die or lifted glue defects in LED manufacturing processes. Usually, surveillance checks like die shear sample tests are carried out to check the die attach quality. Poor die shear strength is one of the indicators for potential lifted die or lifted glue issues. This is not a full-proof detection method. Analyzing the die-sheared specimen will give a clue about the lifted die or lifted glue issue. Figure 3.15 shows an example of a lifted die or glue.

If the die shear test failed and there is no glue remnant on the substrate surface, it indicates poor adhesion of glue on the substrate surface. This shows there could be potential substrate surface contamination. On the other hand, if the die shear failed and the die backside has no glue remnant, then it indicates poor adhesion between glue and die surface. There could be many reasons for such contamination. One of them can be the mylar adhesive that contaminated the die backside. Usually, this is due to die on the mylar stored for a long time in an environment that is not well controlled. The controlled environment is at $25 \pm 3\,^{\circ}\mathrm{C}$ at 50% RH. Another reason that can cause such issues is contamination from the wafer processing processes. Hence, to pinpoint such a root cause, a detailed surface contamination analysis is required, and a thorough investigation is needed.

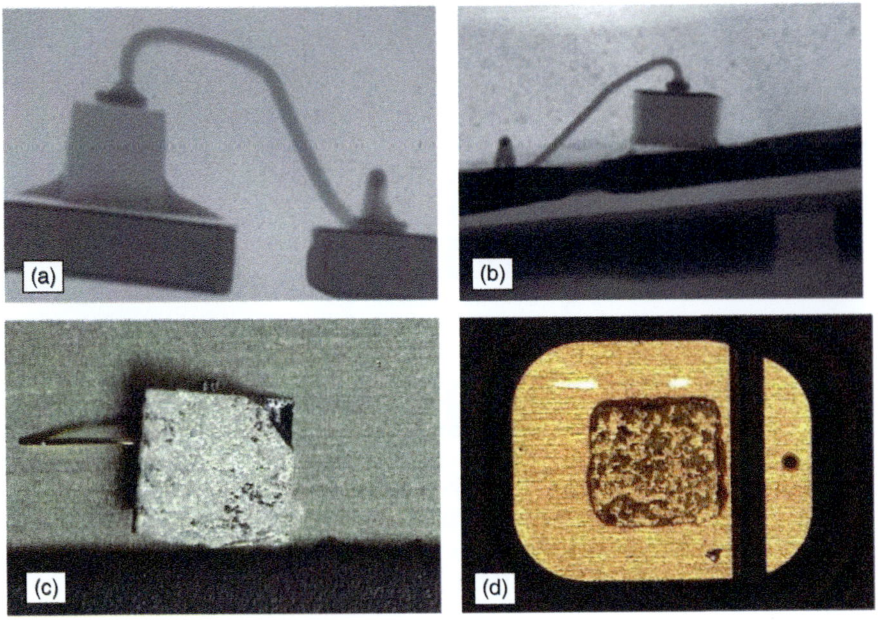

Figure 3.15 (a) Lifted glue LED failure X-ray photo, (b) lifted die LED failure X-ray photo, (c) lifted glue where glue remnant is on the backside of the die, (d) lifted die where glue remnant is at the substrate surface. Source: Courtesy of ams OSRAM GmbH.

Lifted die or lifted glue can also be attributed to package design, which is associated with high stress induced on the die by the package. However, this is a systematic failure, which is the least of the other causes that have been explained earlier.

Lifted die or lifted glue issues can have a very damaging effect on LED applications, as they are not easy to detect at the LED manufacturer level. Many of the failures occur after mounting the LED on the board, or in the worst case, during the end-customer operation level. Given this issue, many attempts to mitigate such failures on the LED manufacturing side have been made. Examples: better process control of the incoming materials at the substrate and die manufacturer sides, robust package design, low-stress encapsulant, better anchor at the substrate surface, and improved detection at LED manufacturer level.

One area to improve detection is the final testing process. At final testing, usually, every single LED is tested for its electrical and optical properties. At the final testing process, if a thermal test was incorporated, the LED with lifted glue or lifted die very likely can be detected. There was one research work carried out by Annaniah et al. [7] using thermal testing methods to detect horizontal cracks. In this work, they have proven that horizontal cracks can be detected. The principle behind this testing method is to use a thermally sensitive parameter, that is forward voltage, to detect these defects [7]. The horizontal crack die has higher thermal resistance compared to no horizontal crack due to the thermal path that has been impeded. Having a higher Rth, the die temperature will increase during operation. This will reduce the forward voltage of the horizontal crack die compared to the die without a crack die. However, this effect is a transient effect and must be measured within a few milliseconds after driving the LED at the operation level based on the findings of Annaniah et al. Hence, to detect this thermal transient voltage, an instrument that is sensitive to this transient voltage is needed [7]. This thermal testing has shown its usefulness in detecting horizontal crack-die. The LED with lifted glue or lifted die also behaves in a similar manner as horizontal crack die LED. Hence, such thermal tests can be facilitated to detect those defects as well.

3.2.3 Wire Interconnect Defects: Impact on LED Electro-optical Quality

The wire interconnect issue has been around since the dawn of the semiconductor industry. In the LED manufacturing industry, especially when using clear encapsulants, such defects are very visible and have a huge impact on electro-optical properties of an LED. Figure 3.16, shows a few common defects in the LED manufacturing industry. Lifted ball, crater on the die, broken wire, and off pad bonding are commonly observed and can be detected during the final test and visual inspection process, especially those LEDs that are encapsulated using a clear compound.

Those defects shown in Figure 3.16 directly influence the LED's electro-optical performance. Lifted ball, broken wire or stitch, and cratered die bond pad will directly affect the interconnection of the LED. The electrical connection is usually

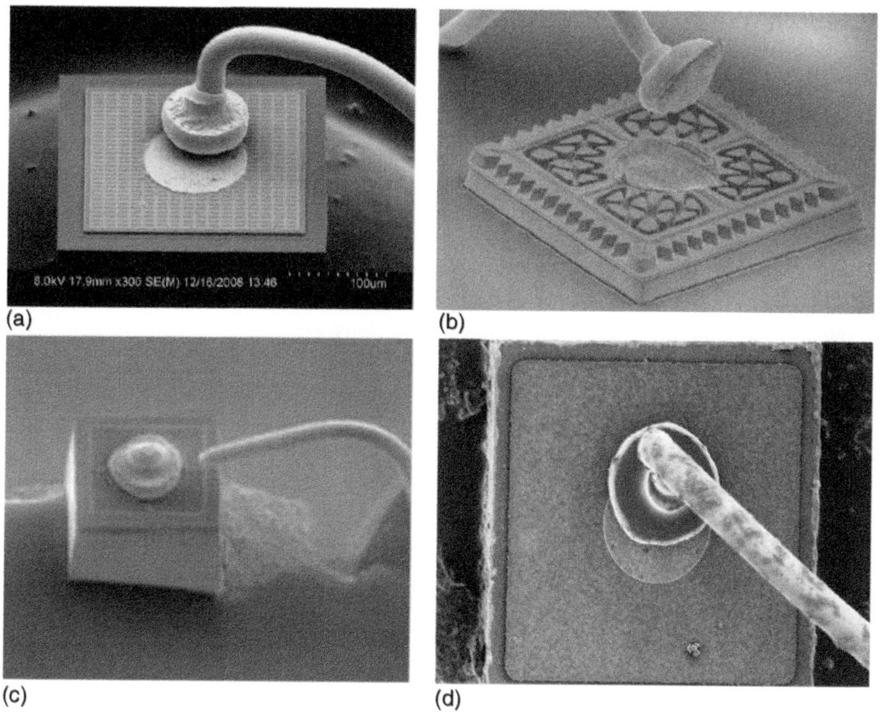

Figure 3.16 (a) Lifted ball, (b) cratering die bond pad, (c) broken wire/stitch, (d) Off Pad. Source: Courtesy of ams OSRAM GmbH.

cut off or intermittent contact. In most cases, the LED ceases to work at zero hours of operation. In some cases, especially with intermittent connections, a wire or cratered die will light up, but the forward voltage will increase as the series resistance increases due to poor connection. These LEDs will ultimately fail after a short operation hour. This is highly risky for application. This wounded LED, in most cases, can be detected using a low-current biasing test. This low-current biasing test (usually <10 μA) is done after high current biasing. High current biasing will aggravate these interconnect defects. In most cases, the LED will be open or show high resistance, thereby directly increasing the voltage at low current. This abnormality can be detected using this low-current test.

3.3 Summary

LED packaging manufacturing is equally important as the other industries in the LED supply chain. Having a competitive advantage in this area is equally important to survive in this competitive LED manufacturing industry.

References

1 Denko, N. (2006). *Nitto Clear Epoxy Moulding Compound NT 8506*. Japan: Nitto Denko Corporation.
2 OSRAM (2020). *Dry pack information*. OSRAM Opto Semiconductor. 1–10.
3 Ching, L.W. and Devarajan, M. (2012). Effect of damaged-chip infrared emitter package on ge substrate. In: *IEEE-ICSE Proceding* (ed. L.W. Ching and M. Devarajan), 532–537. Kuala Lumpur: IEEE.
4 Lu, G., Yang, S., and Huang, Y. (2009). Analysis on failure modes and mechanisms of LED. In: *Reliability, Maintainability and Safety, 2009* (ed. G. Lu, S. Yang, and Y. Huang), 1237–1241. Chengdu: IEEE.
5 Chock, R. (2013). The new genre of application specific LEDs. In: *Strategies in Light* (ed. R. Chock), 2–25. Yokohama, Japan: PenWell.
6 OSRAM (2015). *OSRAM internal quality report*. Bayan Lepas, Malaysia.
7 Annaniah, L., Devarajan, M., and San, T.K. (2017). An investigation on die crack detection using temperature sensitive parameter for high speed LED mass production. *Results in physics* 7: 3882–3891.

4

LED Automotive Lighting Application Technology

The automotive industry has evolved over a century. The innovation in this industry is truly amazing. It has transformed the vehicle's design, style, ecstatic value, human–machine interaction, and user safety to very high level. This somehow pushes all the automotive supporting industries in parallel to meet this demand for innovation. One of the industries that has significantly benefitted is automotive lighting.

Automotive lighting has provoked human ingenuity to a next level, where many amazing ideas, concepts, technologies, designs, and science have been discovered. Its evolution over the past century has been mind-boggling. Driving much of that evolution in light sources and their luminance. Growing standardization around electrical light sources gives a distinct opportunity for LED lighting technology to emerge as a major benefiter. LED, as we all know, can be designed in many possible designs to fit the automotive lighting application.

The lighting system for automotive vehicles plays a critical role in safety of both the driver and other road users **to see** and **to be seen**. The risk of having a car accident at night is three times higher than that during the day. Automobile makers and component manufacturers are consistently developing new technologies to offer a wider and longer field of vision to improve anticipation under driving conditions.

What are automotive lighting applications? As mentioned earlier, the automotive lighting system basically covers two aspects: (i) **to see** and (ii) **to be seen**. The lighting "**to see**" is called **lighting**, which covers light to project to anticipate the road ahead (bends, pedestrians, road signs, threes, animal crossings, etc.). The better you can see the safer you are. On the other hand, the light "**to be seen**" is called **signaling** light. This light is to allow light to be seen by others – drivers and pedestrians. Being seen allows others to better detect and anticipate your movements. In a car, there are many types of lights. Figure 4.1 illustrates the automotive lighting system in general terms.

Both **lighting** and **signaling** light are specified and designed according to stringent regulations. All the lighting is governed by European Commission regulations for automotive lighting systems, as classified in Table 4.1.

LED Packaging Technologies: Design, Manufacture, and Applications, First Edition.
Luruthudass Annaniah, Mohamed Salleh M. Saheed, and Rajan Jose.

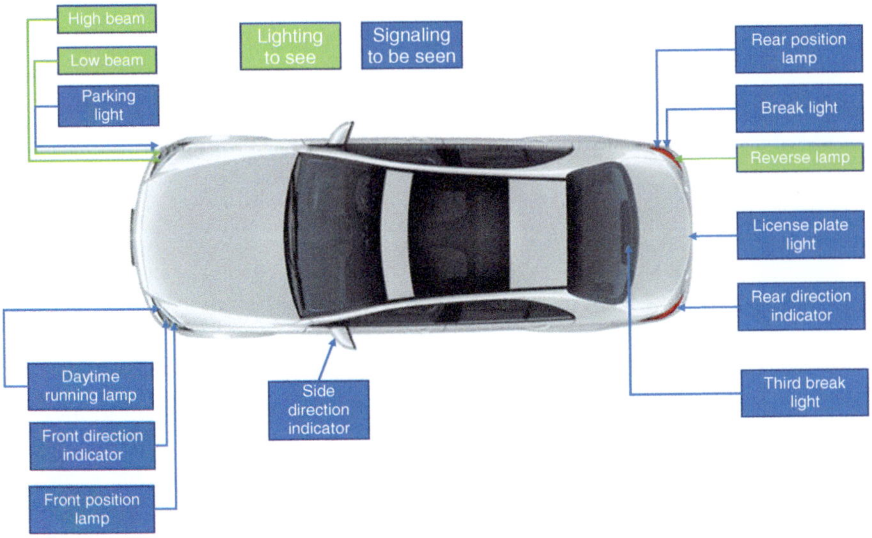

Figure 4.1 Automotive lighting system that shows various lights that cover lights "**to be seen**" and "**to see.**"

Table 4.1 European Commission regulation for automotive lighting systems.

R37	Light sources – filament lamps
R99	Light sources – HED lamps
R112	Low and high beams – for halogen lamps and LED sources
R98	Low and high beams – HID lamp sources
R19	Beams – front fog lamps
R87	Daytime running lamps
R48	Set on the car
R123	AFS beam – halogen, xenon, and LED light sources

4.1 Basic Science of Light for Automotive – The Photometric

Before we deep dive into the automotive lighting system, we must understand the basics of photometrics. Photometry is the science concerned with measuring human visual response to light. Here are a few basics that are good to know prior jumping into the lighting system.

4.1.1 Light Intensity

The intensity is the luminous energy from a lighting source in a specific direction. The symbol for light intensity is I, and the unit used for lighting intensity measurement is **candela (cd)**. One candela represents the light intensity of one candle [1].

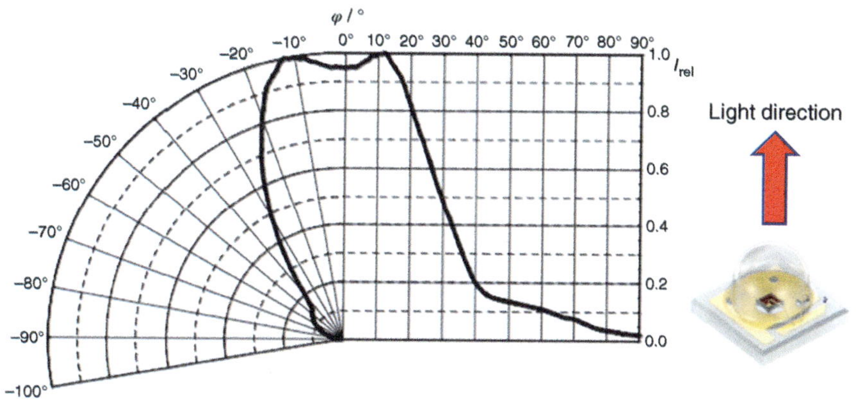

Figure 4.2 Radiation pattern of an LED – OSLONSX – LA CN5M. Source: Courtesy of ams OSRAM GmbH.

The light intensity varies depending on which direction the light source is being viewed from. Figure 4.2 shows an OSRAM LED – OSLON®SX LA CN5M radiation pattern [2]. The figure shows at 0° the intensity at 95% and when it is seen directly at 10° the intensity at max. The intensity drops drastically if we view at an angle greater than 10°. The light intensity of all automotive lighting systems is regulated. An example of stop light is 60 cd. Parking light is 2 cd, and position light is 4 cd. In Section 4.3, this lighting will be further elaborated.

4.1.2 Luminous Flux

The luminous flux is the complete light output radiated from a light source, for example, automotive LED lamp. The symbol used is **Ø**, and the unit used to measure it is **lumen (lm)**. The lumen unit is defined based on the human eye's sensitivity curve. Figure 4.3 shows how a human eye perceives a light source depending on its color composition at daytime and nighttime. At daytime called **photopic** vision

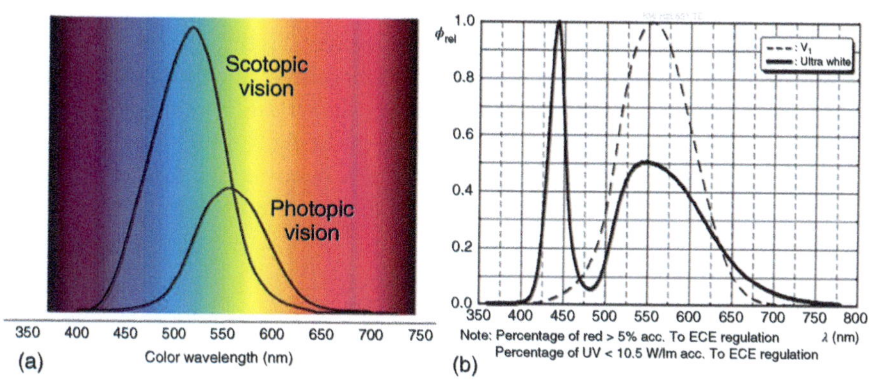

Figure 4.3 Spectral sensitivity of human eyes. Source: Courtesy of ams OSRAM GmbH [3, 4].

and at nighttime called **scotopic** vision [1, 5]. The human vision sensitivity peaks at blue color (507 nm) wavelength at nighttime, and on the other hand, at daytime sensitivity peaks at green color (555 nm) [6].

The theoretical and ideal light source for a human eye would have a spectral decomposition as shown in Figure 4.3a,b. This curve (**scotopic** vision) comes close to an LED light source (Ultra White LED), as illustrated in Figure 4.3b.

4.1.3 Illuminance

The **luminous flux** from a light source (**lumens**) falling on a unit surface (m^2) is called illuminance. The symbol used for illuminance is E, and the unit for measurement is **lumen/m²** also called **lux**. A good example of illuminance application is the automotive headlamp. Illuminance levels on the road are specified according to the automotive standards, and they depend on the type of light sources, such as LEDs or other lamps. The illuminance level does not make allowance for light reflected off the surface. The reflected light is called the **luminance**. This depends on the nature of the surface and its color. Roads and road marking are specified in standards that set minimum reflection levels [1, 5]. A good example of luminance application is day running light (DRL).

4.1.4 Luminance

The **luminance** measures the reflected light from a surface in a given direction. The surface can itself be light emitting, transmitting, or reflecting light from another light source. The **luminance** is perceived brightness; it is used to categorize the light sources (example LED lamps), lighting systems (example headlights), and any surface that is bright like road signs. The **luminance** is related to a luminous intensity emitted per unit of area from a surface in a specific direction. The symbol used for **luminance** is L, and the unit used for measuring it is **candela per m²** [5].

4.1.5 Luminous Efficacy

Luminous efficacy is a figure of merit for light source. It is the proportion of visible radiation over the entire radiation emitted by a light source. The **luminous efficacy** is expressed in **lumens per watt**; it is the ratio between the total amount of electrical power (W) it consumes. The higher the efficacy, the better it is in terms of electrical power consumption. An LED efficacy of >100 lm/W has overpassed the xenon lighting technology. This makes LED an ideal replacement for most automotive lighting systems [5].

4.2 Lighting – Light Projection "To See"

As mentioned earlier in the chapter, the automotive lighting system can be divided into two subfamilies that is lighting (**to see**) and signaling (**to be seen**). In this section, we focus on the lighting **"to see,"** that is, the headlamps. The headlamps have evolved over many decades. In this section, we will cover the headlamp functions and some of the developments related to headlamps like adaptive front-lighting systems, future headlamp technology, and LED headlamp thermal management.

4.2.1 Headlamp

A good headlamp function provides optimum visibility with minimum inconvenience to other road users. The headlamp beams are standard patterns that set the light projection levels and limits on the road. The headlamp contains several beams as shown below:

a. Low beam, also called dipped beam
b. High beam, also called full beam
c. Fog beam.

These beams are defined according to their shape and performance such as:

a. The **"width"** area, where illuminance covers the whole road width in the driver's near field of view.
b. The **"comfort"** area, which corresponds to the driver's main vision zone.
c. The **"range"** area, where the optimum illuminance levels occur far on the road ahead.

The headlamp is categorized into a few categories: low beam, high beam, and fog beam. The low beam, as illustrated in Figure 4.4a, is intended for use whenever other vehicles are present ahead. Figure 4.4b illustrates the location of the low beam. Most of the automobiles on the market today have opted for LED light sources, as shown in Figure 4.4c. It offers long lifetime, small size, and very high efficacy.

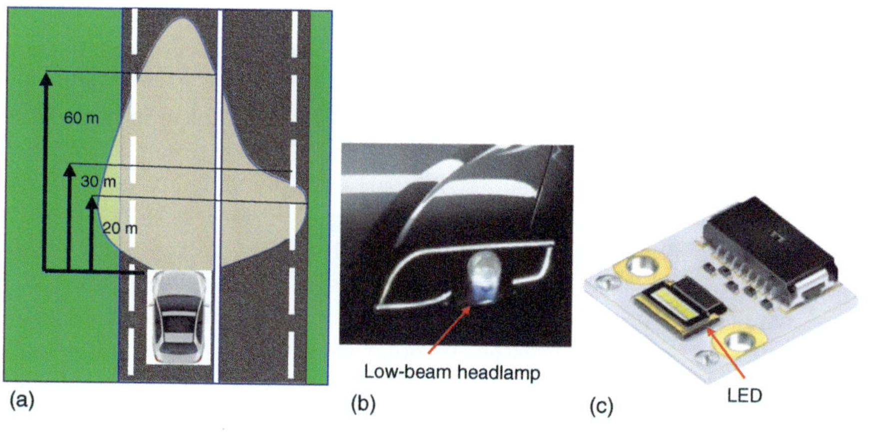

Figure 4.4 (a) General description of a low-beam lamp on a road, (b) low-beam headlamp, (c) OSTAR Headlamp Pro. Source: Courtesy of ams OSRAM GmbH.

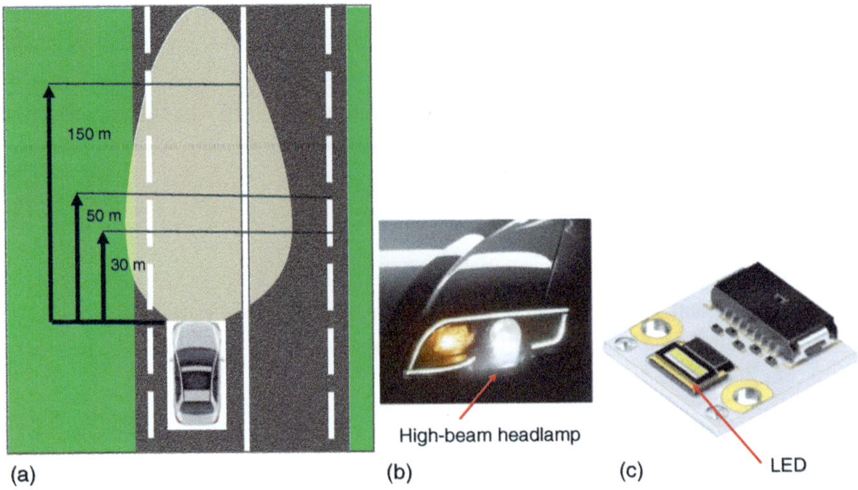

Figure 4.5 General description of (a) high-beam lamp on the road, (b) high-beam headlamp, (c) OSTAR Headlamp Pro. Source: Courtesy of ams OSRAM GmbH.

The low beam as illustrated in Figure 4.4a, has a sharp, asymmetric cut off that prevent significant amounts of light from being emitted onto the eyes of the drivers of preceding or oncoming cars [7, 8].

The high beam, as shown in Figure 4.5a, on the other hand, provides a bright, center-weighted distribution of light with no control of light directed toward other road users' eyes. This high beam will dazzle other drivers coming from opposite direction. It is only suitable for use when alone on the road or median highways [7, 8]. The high beam, as illustrated in Figure 4.5b, is located in the same compartment as the low beam but without an asymmetric cutoff. This high beam light source in most of today's automobiles has opted for an LED light source, as the brightness and efficacy of an LED light source have superseded xenon or halogen lamps. An example of the LED headlamp light source is OSTAR® Headlamp Pro, as shown in Figure 4.5c.

The fog beam as shown in Figure 4.6a, on the other hand, provides a wide, bar-shaped beam of light with a sharp cutoff at a range of 25 m. Fog lights are designed to be used at low speed in foggy poor visibility situations or in heavy mist, snow, and other poor visibility condition on the road. They are different from daytime running lights or low-beam lamps. They are extra pair of lights mounted low on the vehicle front-lighting system, as shown in Figure 4.6b [7, 8]. And most of the fog lamps have been converted into LED lamps lately due to similar reasons of high efficacy and durability. They are fixed into fog lamps, as illustrated in Figure 4.6c.

One of the challenges of nighttime driving is dazzle issue. Many accidents are associated with this dazzle issue. In order to solve the dazzle issue, many automakers opt for adaptive front-lighting systems (AFS). In this system, the high beam and low beam are automatically controlled by the car in the event there is a car approaching opposite the driveway. In Section 4.2.2, this AFS will be further elaborated.

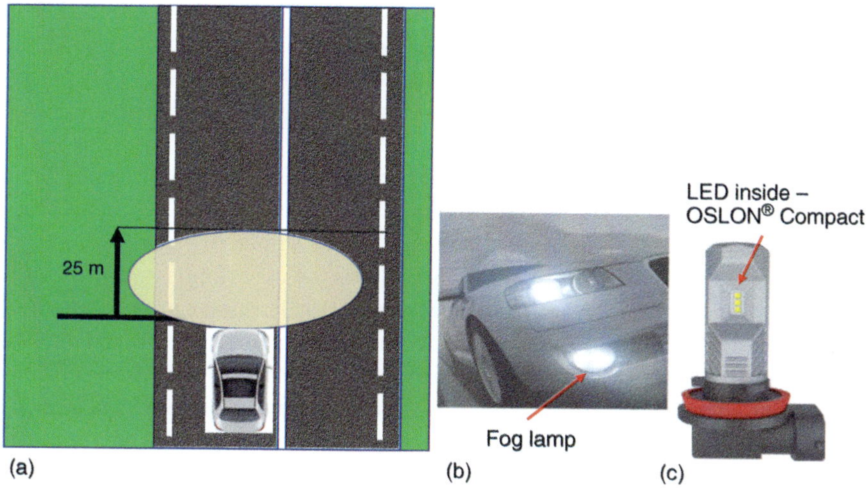

Figure 4.6 General description of (a) fog beam lamp on the road, (b) fog lamp, (c) OSLON Compact. Source: Courtesy of ams OSRAM GmbH.

4.2.2 Adaptive Front-Lighting System – Headlamp

Function of AFS is to optimize the distribution of light from the headlights according to driving circumstances. Depending on vehicle speed and steering input, the system points the low-beam headlights in the direction the driver intends to travel. It is one of the most important factors in mitigating driver fatigue and increasing safety during night driving by providing a well-illuminated field of view [9]. AFS is designed to reduce the risk of accidents. However, the system has its limitations, and no safety system or combination of such systems can prevent all accidents. These systems are not a replacement for safe and attentive driving [10].

AFS was first introduced by Citroen DS in the 1960s with a primitive version of headlamps. It uses a mechanical system where the inner pair of lights swiveled with front wheels, enabling them to see around the corners. Ever since then, the AFS has evolved. Figure 4.7 shows the roadmap of AFS evolution.

With the advances in design and technology, the ability to control headlamps and the availability of robust electrical actuators have allowed designers to mechanically

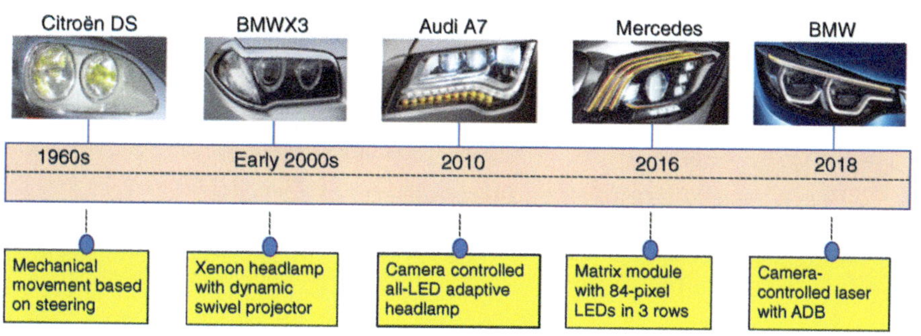

Figure 4.7 Evolution of adaptive front light system.

move light, or shade it, at their will. Technically, the simplest level of adaptation is auto-leveling. These were the advanced actuator-driven units. This auto-leveling system originates from work done by the Eureka project by European auto manufacturers, lamp makers, and light source makers to develop an intelligent lighting system (ILS) to improve the safety and comfort of nighttime driving and reduce traffic accidents related to glares [8]. At the same time in Japan, Koito was also working on a similar system for Japanese automakers [11].

This auto-leveling headlamp system in simple term is shown in Figure 4.8. When a car accelerates quickly, the weight of the body causes the car to angle upward, directing a glare at oncoming vehicles. Conversely, rapid deceleration causes a nosedive, reducing the forward lighting performance. A similar change occurs with changes in the number of passengers or loads being carried. To manage the headlamp, axis, the dynamic auto-leveling headlamp system, as shown in Figure 4.8, has a vehicle speed sensor and two-axle sensor attached to both front and back axles, which provide the data for maintaining the ideal headlamp beam axis.

This dynamic auto-leveling headlamp system was adopted for use in Mercedes Benz and BMW car models in 1995. Japanese carmakers also followed this system for nearly similar period of time [11].

In the middle of 2006, the first car equipped with an intelligent AFS headlamp went into series production and has generated a very positive reaction in the press all over the world. The different AFS beam patterns in this car are generated by the xenon AFX projector module, which works in combination with a swiveling actuator system.

One great step in AFS innovation was the introduction of an LED light source in 2010 for headlamps, when Audi A7 introduced camera-controlled all-LED adaptive headlamps. The performance level of this light source technology is well-accepted when compared with that of high-intensity discharge (HID) headlamp system. One of the most significant differences between an HID bulb and an LED is the difference in lighting performance, especially regarding the luminance and luminous flux output. In the early days, the LED brightness was not able to match the brightness level of HID bulb performance. HID lamp typically has a luminance of 90 cd/mm^2 and a luminous flux of around 3300 lm generated by the arc. On the other hand, for LED, luminance is up to 20 cd/mm^2 and the luminous flux is about 300 lm (for a multi-chip LED with five chips) [8]. However, LED performance has improved drastically ever since then. In 2013, Audi released the A8 series, which

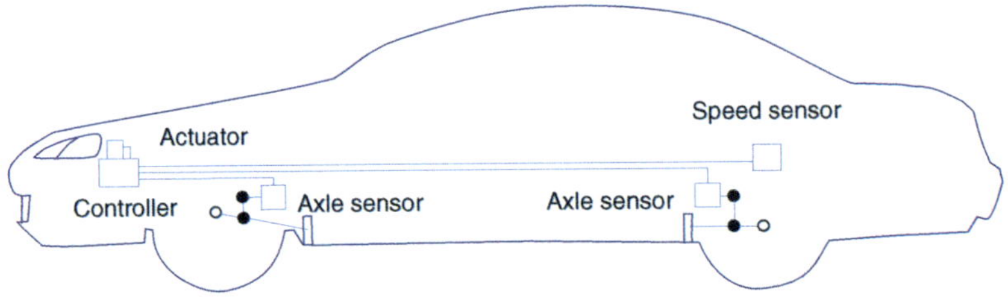

Figure 4.8 Dynamic auto-leveling headlamp system. Source: Kobayashi [11], SAE International.

came with Matrix LED headlamp that was one step ahead of the adaptive headlamp technology. While increasing the safety of the driver and pedestrians, the company developed a more robust and precise lighting system with added functionalities [12]. The headlights would not only assist in illuminating side roads while parking but also reduce the glare from the oncoming traffic and allow the driver to always keep the high beams ON.

The heart of the Matrix LED headlights is the mechanic-free glare-free high beam. This allows the driver to travel in his vehicle on a permanent high beam without the risk of dazzling oncoming traffic or any preceding vehicles. Employing a camera, oncoming and preceding traffic are detected and then, by the shutting down or dimming of individual LEDs, the field of high beam light distribution in real-time are blanked out toward these vehicles. The implementation of Matrix technology allows, for the first time, several tunnels to open simultaneously. One example of this is the scenario, as illustrated in Figure 4.9a, where several oncoming vehicles are driving one behind the other. While these are "masked out," the high beam continues to illuminate all the areas between the vehicles and to the right and left of them at full power. As soon as no vehicle is any longer in the driver's field of vision, the system once again reverts to full high beam lighting, as shown in Figure 4.9b. In addition to the specific masking out of other vehicles, the light cone of the Matrix high beam also adapts to the driving situation, for instance, in the case of navigating bends when the dynamic bending light function is required. In such a situation, the intensity of the light cone can be varied on the sides or it can be focused on the middle of the road. Consequently, the driver's visibility at night improves dramatically while, at the same time, the risk of dazzling oncoming traffic is eliminated [13].

Such a function is technically possible thanks to the splitting up of the high beam into five reflectors, each one having a chip containing five LEDs, as shown in Figure 4.10. For the first time, the lighting expert, HELLA, has now succeeded in operating every LED on the 5-segment chip separately, whereby a total of 25 LEDs per headlight can be operated at full power or lowered as and when required.

The Matrix headlight has further evolved. In today's matrix-beam units like EVIYOS of OSRAM, single LEDs illuminate defined sections of the beam pattern and are controlled individually by an electronic control unit wired to camera systems. Besides vehicles, the camera can identify pedestrians and animals close

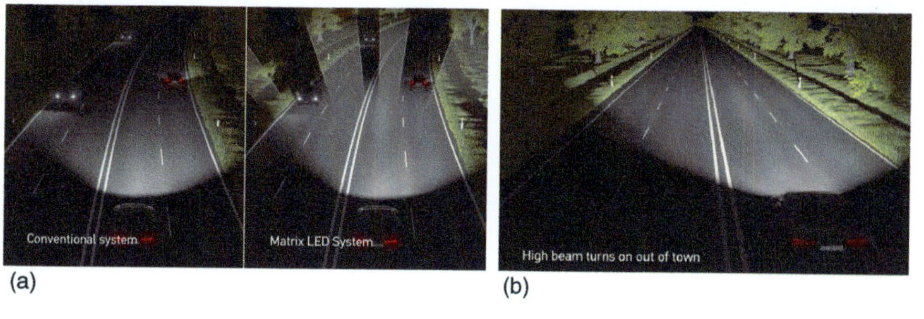

Figure 4.9 Headlamp comparison between the Conventional System and Matrix LED System when present of car and no car on the road. Source: Image from FORVIA Hella [13].

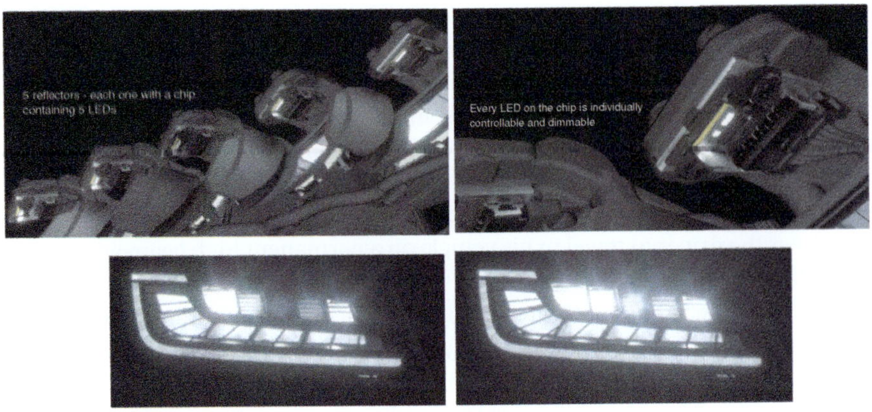

Figure 4.10 Matrix LED headlights. Source: Image from FORVIA Hella [13].

to or on the road, and then notify the driver (using a marker light) or un-glare the headlights. In addition to its significantly lower power consumption, the LED system is lighter than the xenon headlight or HID headlight and, above all, less susceptible to vibration. Besides, matrix-beam LED units can provide other benefits [14]. EVIYOS-based matrix-lighting system will be further elaborated in a later chapter.

4.2.3 Optical Concept Automotive Front Lighting – Headlamp

Forward lighting plays an important role in vehicle safety. The generation of light functions with conventional headlamps is based on two different technical concepts that are refraction and reflection technology. Figure 4.11a,b shows the headlamps

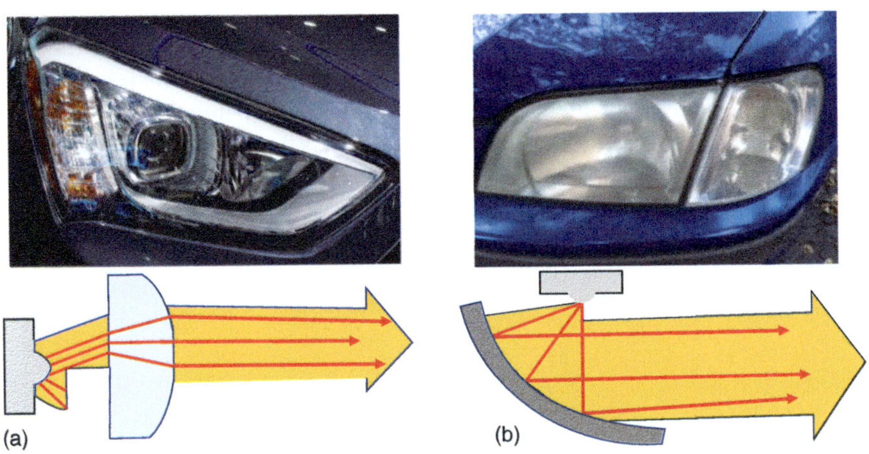

Figure 4.11 (a) and (b) respectively show the headlamps refraction and reflection technology concepts. Source: Courtesy of ams OSRAM GmbH; Wördenweber et al. [7]/Springer Nature.

refraction and reflection technology concept, respectively. The reflection systems stand out due to large reflectors behind a patterned or clear lens. On the other hand, the refraction systems have a small light aperture with a characteristic lens, which can be enlarged by brightening the surrounding area. Reflection systems require less installation depth compared to refraction systems but are wider and higher [7].

4.2.4 Future of LED Headlamp Technology

LED in front lighting has been a common feature in the latest high-end cars. Digital Matrix LED beam headlamps have become the preferred solution for headlamp makers. The conventional Matrix LED beam headlamps are mostly made of multi-discrete LED or multichip LED package that are well accepted. Moving forward, the Matrix LED headlamp has taken further steps of miniaturization and compacting the LED matrix. OSRAM has developed a new concept of LED matrix solution for headlamps, called EVIYOS®. EVIYOS is the first smart and controllable LED that combines a micro-structured LED chip with an IC capable of generating pixelated and individually controlled light in just one device. This combination means that the EVIYOS, which has 1024 pixels plus one IC, is accommodated in a single package with a footprint of 4 mm × 4 mm as shown in Figure 4.12.

The light source has a minimum luminous flux per pixel of 3 lm at 11 mA. This corresponds to roughly 3000 lm per LED. This light source is twice as much as the luminous efficacy of a halogen headlamp. The comparatively high contrast of at least 300 : 1 also contributes to the high quality of differentiated illumination.

In comparison to the previous LED matrix solution that is based on individual LEDs, this EVIYOS has much higher resolution. The current LED headlight works with multiple LEDs, which is less than 100 LEDs. The resolution of the new lighting (EVIYOS LED) is about 10 times higher. A higher resolution enables more precise masking of smoother beam shapes. The 1024 pixels are automatically switched on or off at a very fine resolution depending on the traffic situation. As shown in Figure 4.13, an example of a digital matrix headlight where the car driver does not have to switch in between high and low beam as this is automatically controlled by the IC embedded in this package.

Figure 4.12 OSRAM EVIYOS LED provides smart lighting with more than 1024 chips of individually controlled pixels. Source: Courtesy of ams OSRAM GmbH.

Figure 4.13 Digital matrix headline is automatically switching on or off at a very fine resolution depending on the traffic situation. Source: REC Anything [15].

These units combine the principles underlying matrix-beam systems with the imaging optics and resolution of a video projector. Whereas a matrix beam with up to a hundred LEDs breaks the light pattern down into small, angular segments specifically designed for their purpose, a pixel light can create any light pattern desired out of thousands of pixels, in the same manner that a video projector composes an image. Thus, for example, a pixel light can write words or display pictures on the road to interact with drivers or pedestrians. The main benefit of these lights is expected to be their unrivaled ability to adapt to any driving situation and provide optimum lighting [13].

4.2.5 LED Headlamp Thermal Management

The LED is not a thermal light source. An LED lamp is no longer an optical system that must withstand the heat of the thermal light source. It is an "opto-thermo-electro-mechanical" system that must manage the balance between each of the constituents within the adverse conditions of a highly visible, hot, dirty, shaken, stone-pecked, and chemical-laden environment. This is an opportunity and adversity all at the same time.

The thermal function is the secondary function, which strongly determines the design. Overheating of the LED is a big concern and limits the power put through the diodes. Considerable work is therefore going into the management of junction temperatures and thermal conduction coefficients of the light source. The headlamp must operate over a range of temperatures from −40 °C to, in some cases, over 125 °C. Two main thermal problems need to be solved in the headlamp: (a) stopping the light source from overheating and (b) defrosting or deicing the headlamp lens. Most designs contain ample passive heat sinks. For a more compact headlamp, active cooling using fans or heat pipes is considered. The continued development of the LED light source is slowly making the integration task easier. The temperature resistance of LEDs is improving. The color tolerances in the production of high-power white LEDs are becoming more and more accurate, and solutions to handle the variation in forward voltage and flux are well managed.

4.3 Signaling – Lights That Are "To Be Seen"

In automotive signaling lights, both automotive front lighting (AFL) and automotive rear lighting (ARL), as illustrated in Figure 4.14a,b, are very important not only for their functions but also to protect the aesthetic value of a car.

AFL has long been evolved, from just a bare incandescent light source to a complex LED light. Its lifetime also increased in accordance with the LED's performance.

The forces that are driving the evolution of AFL and ARL, as illustrated in Figure 4.15, are safety, reliability, styling, effectiveness, serviceability, and adjustability of the front lighting. In developing AFL and ARL solutions, the developer must consider these factors to meet the intended end application and comply with regulations stipulated in respected regions.

One of the main driving factors of the automotive front- or rear-lighting system is the styling, which has become a major selling point of the automotive business, particularly in high-end cars. Most of the large automakers are combining the styling and the core functions of the AFL and ARL to provide an awesome aesthetic

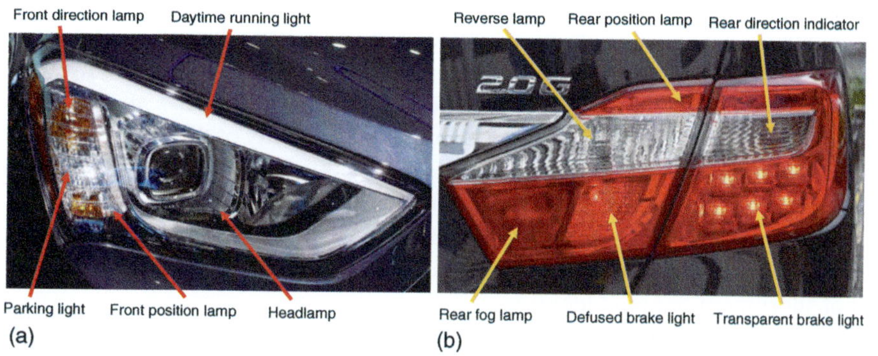

Figure 4.14 (a) Front lighting – consists of headlamp, DRL, front direction lamp, parking light, and front position lamp. Source: Courtesy of ams OSRAM GmbH. (b) Rear Lighting – consists of brake lights, rear direction indicator, rear position lamp, reverse lamp, and fog lamp.

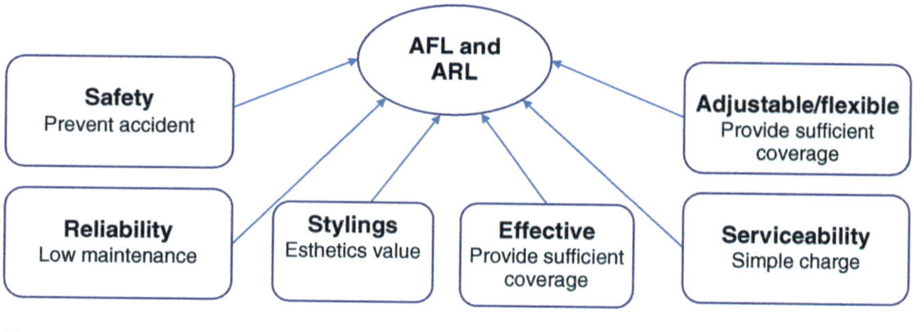

Figure 4.15 Driving forces of the automotive front lighting (AFL) and automotive rear lighting (ARL) systems.

value to the end customer. In Sections 4.3.1–4.3.3, these lightings will be further elaborated.

4.3.1 AFL - Day Running Light

DRL was originally made popular in parts of the world where daylight can often be dim and short in duration. It was first introduced in the 1970s as a safety measure and soon has become very popular in countries located further north, where there is less light (especially in winter). It makes sense, then, that countries such as Sweden, Norway, Iceland, Denmark, and Canada were among the first to require DRLs on all their vehicles. In these countries, multiple studies conducted since the 1970s have shown that the addition of daytime running lights does reduce accidents. In the United States, there was resistance to the use of DRL in the early days. Eventually, the government regulation softened and allowed cars with DRLs to be driven on US roads as the benefits speak for themselves in other developed countries.

DRL works great, to identify the car using them to other cars on the road. Unlike headlights, they are not intended to illuminate the road ahead. DRL is the car light in the "**to be seen**" category. They are simply dim lights at the front of the vehicle, as shown in Figure 4.16. The DRL is situated right below or sometimes above the headlamps. It has been a common feature in the AFL nowadays in most parts of the world. The DRL is regulated in accordance with ECE R87. The intensity along the H and V axis, between 20° to the left and 20° to the right and between 10° up and 10° down, as shown in Figure 4.16b. The luminous intensity of the light emitted by each daytime running lamp shall not be less than 400 cd in the axis of reference. It shall not exceed 1500 cd in any direction the DRL is visible [17].

Almost all the new cars now have this DRL feature in their vehicles. This DRL system is automatic. As time went on, many automakers have adopted DRLs as a sort of adornment or brand identifier.

DRL not only has aesthetic value, but also its primary function is to provide essential safety to road users, especially oncoming cars, pedestrians, and other road users. Fundamentally, today's DRLs are bright, low-powered lights using LED technology, which can operate all the time that your car is running.

Daytime running light (DRL)

(a)

H (horizontal), V (vertical)

Daytime running lamps	20°	10°	5°	0°	5°	10°	20°
UP							
10°			80	80	80		
5°	40	80		280		80	40
0°	100	280	360	400 cd	360	280	100
5°	40	80		280		80	40
DOWN							

(b)

Figure 4.16 (a) Audi R8 daytime running light and front turn signal light. Source: Courtesy of ams OSRAM GmbH; Morkos et al. [16]/SAE International. (b) ECE R87 (daytime running lamps), the luminous intensity of light emitted by each lamp shall not be less than 400 cd in the axis of reference.

4.3.2 ARL – Signaling Lights

At the rear and the side of the car, some signal lamps show the other traffic participants what your car is doing, or what you are about to do. As shown in Figure 4.14b the rear position lamp, brake light, rear turn indicator, and reversing lamp are usually combined into a cluster referred to as the rear combination lamps or tail lamp. Other lamps, such as the third brake light or the side direction indicator, are typically single-function lamps. For each lamp, a large variety of different styles and designs exists. In this section, we will look at ultra-thin lamps with a mounting depth that is measured in millimeters.

One major advantage of LEDs in signal applications is freedom of design. LEDs for signaling created lighting quality issues. When LEDs displace conventional bulbs, traditional considerations of how to build good signals are discarded. The small package with continually growing source lumens brought high luminance into the world of signal applications – it is now possible to dazzle other drivers with excessive glare from brake lights. Regulations today require automotive lights to comply with photometric regulations. One example of a rear light that is stipulated in ECE R07 regulation is shown in Figure 4.17. As to the minimum values, they are defined in Figure 4.17a,b for a rear position of 4 cd and for brake lights of 60 cd, respectively, with respect to the specific directions [18].

Light guide optic designs, as illustrated in Figure 4.18, in the Mercedes CLA200 tail lamp, create a distinctive appearance and a sense of depth from a relatively flat

H

Rear position lamps	20°	10°	5°	0°	5°	10°	20°
UP							
10°			0.8		0.8		
5°	0.4	0.8		2.8		0.8	0.4
V 0°		1.4	3.6	4 cd	3.6	1.4	
5°	0.4	0.8		2.8		0.8	0.4
10°			0.8		0.8		
DOWN							

(a)

H

Brake light	20°	10°	5°	0°	5°	10°	20°
UP							
10°			12		12		
5°	6	12		42		12	6
V 0°		21	54	60 cd	54	21	
5°	6	12		42		12	6
10°							
DOWN							

(b)

Figure 4.17 The photometric testing requirement of the ECE R07 standard. (a) Rear position lamps and (b) brake lights. Source: Le et al. [18]/MDPI/Public Domain CC BY 4.0.

Figure 4.18 The Mercedes CLC200 tail lamp, while a very flat design, gets a "tunnel look" and a sense of depth from its LEDs and their multiple reflections.

footprint and demonstrate what has become possible with LEDs and state-of-the-art design capabilities.

The light sources for ultra-thin automotive signal lights and brake lights are made of LED or OLED. Their advantages in comparison to a bulb, like quick response time and long lifetime, also apply here. Besides these general advantages, an ultra-thin signal lamp offers the car manufacturer special features. A strongly reduced mounting depth of the lamp allows:

(a) a higher usable volume inside the boot of the car.
(b) weight reduction (less material is required for the lamp).
(c) higher stability of the car body (no need to cut out holes for the lamp).
(d) a cost reduction in the manufacturing of the car body (it is easier and cheaper just to deep-draw the sheet metal instead of cutting a hole in the body, then rebuilding and connecting a deeper-lying metal sheet surface onto the outer body afterward).

4.3.3 Optic Concepts of Signaling Light "To Be Seen"

A car signal lamp and break lights consist essentially of three groups of optics concepts:

(a) reflective optics
(b) refractive optics
(c) light guide optics

In Sections 4.3.3.1–4.3.3.2, we will explore each of these optic concepts.

4.3.3.1 Reflective and Refractive Optics

A conventional car signal lamp with an incandescent lamp, as illustrated in Figure 4.19a, a **reflective optics** concepts that consists essentially of three assembly groups: the **bulb** positioned relative to the optical system of the lamp **reflector** and **cover lens**. The housing contains reflectors, which are generally molded into one, and the cover lens at the exterior part of tail lamp, is usually responsible for the distribution of the light that has **additional optical** structures to disperse the light. The conventional car signal tail is generally much thicker than an LED-based tail lamp [19].

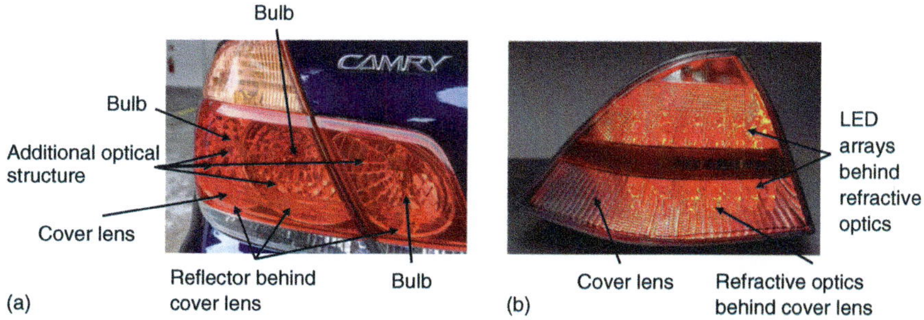

Figure 4.19 Comparison between conventional tail lamps using reflective optics and LED-based tail lamp using refractive optics concepts, (a) incandescent rear combination lamps [7, 19]. (b) Lighted automotive LED tail lamps showing the cover lens dispersing the light. Source: Image from FORVIA Hella.

The LED-based tail lamps uses **refractive optics** concepts as illustrated in Figure 4.19b is a simpler design. It uses LED, integrate corresponding optical components such as **refractive optics**, and **cover lens** [19]. The refractive optics is in between the LED and cover lens, as illustrated in Figure 4.19b collects the light from the LEDs and disperses the light. However, this direct lighting with arrays of LEDs and refractive optics often causes glare problems in tail lamps [18]. This is where the light guide optics come in very handy.

4.3.3.2 Light Guide Optics

Advancements in computer simulation and optic design became easier to achieve. Computer optic simulation permits the design of complex light guides based on LED. This technology can be used to create homogeneously illuminated surface appearance. In order to further understand this light guide optic, we have to understand the science behind this technology.

Light guiding is based on total reflection occurring at the transition interface between the optically thicker and the optically thinner medium. The limiting angle up to which light guiding is possible is determined by the difference between the refractive indices of both materials shown in Figure 4.20. If beams meet the interface at a steeper angle, they are outcoupled.

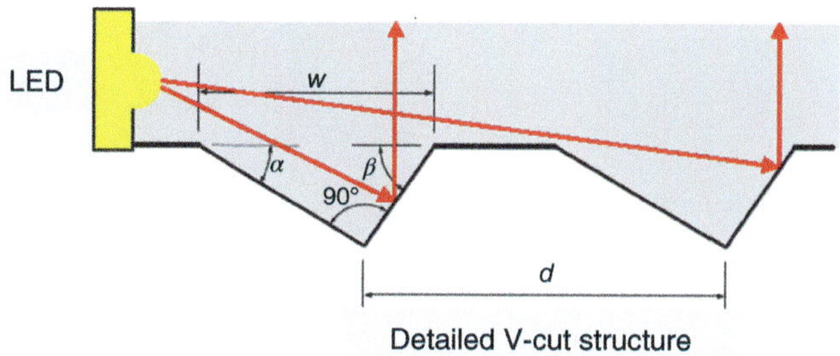

Figure 4.20 V-cut structure of light guide showing the limiting angle for light refraction.

Light guides offer great opportunities to design attractive exterior lamps, e.g. front position lights, tail lamps, or sidelights. Light guides are used for signature lighting to highlight important styling lines, contours around lighting chambers, or the edges of a lamp, as illustrated in Figure 4.21.

The light from the LED light source is coupled into an entrance surface of the light guide made of transparent material, and it is guided through the light guide by total internal reflection (TIR). The refractive index of the surrounding material must be lower than the refractive index of the light guide material. In the case of automotive application, the surrounding material is typically air. To decouple light from the light guide, a prismatic structure is used, as illustrated in Figure 4.22a,b, respectively. There are two concepts of the light guide using LEDs. One is using a direct light coupling optic, as illustrated in Figure 4.22a. Here the LED is close to the light guide. In this light guide optics concept, the LEDs used are the narrow radiation pattern type. Usually, this optic concept is thinner and consumes less space. Ideal for rear lighting and interior lighting. The other light guide optics concept is the collimated light coupling optic. Here the LED is close to the light guide with collimated optics, as shown in Figure 4.22b). For this type of light guide optics concept, the LEDs used are the ones with a wide radiation pattern. This light guide optic is relatively thicker than direct light guide optic as a comparison.

Today, most of the light sources are LEDs due to their low heat emission and small dimensions. It is possible to place LEDs very close to the plastic light guide and

Figure 4.21 Light guide for automotive tail lighting using a light guide optics concept. Source: Image from FORVIA Hella.

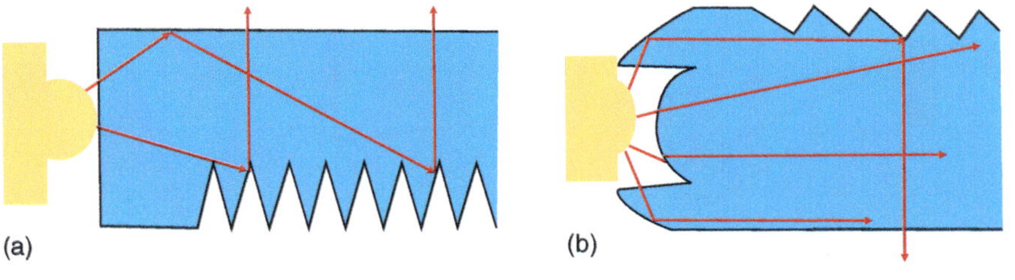

Figure 4.22 Light guide using LED as a light source [7]. (a) Direct light coupling optics. LED close to light guide. Usually for LED with narrow radiation pattern. (b) Collimated light coupling optics. LED close to light guide with collimated optic feature. Usually for LED with wide radiation pattern. Source: Wördenweber et al. [7]/Springer Nature.

achieve efficient light coupling. As mentioned earlier, depending on the radiation pattern of the LED used, the light can either be coupled directly into the light guide or collimated by the input surface of the light guide, as illustrated in Figure 4.22. The light rays from the LED that reach the prismatic structure either directly from LED or reflected from the upper wall of the light guide will hit the prism and be reflected in the direction set by the prism as shown in Figure 4.22a,b, respectively.

One of the key unique appearances of light guide is its uniform light appearance. Light uniformity is influenced by many parameters, e.g. the radiation pattern of the light source used, the geometry of the prismatic structure, the material used, or the scattering losses of light, etc. Combined with the radiation pattern and distance of the LED light source, this usually leads to an exponential decrease of the light flux inside the light guide as the light passes through the light guide. This situation is shown in the following Figure 4.23, where the dependence of the viewing angle on the distance is shown along with the light guide. This clearly can be seen optically on the actual rear light of the Mercedes CLA200, as shown in Figure 4.24a, where the light flux inside the light guide changes as the light passes through the light guide.

The shape of the prismatic structure influences the direction of the decoupled rays from the light guide and tailors the spreading of the light beam. Figure 4.24a,b shows an example of a typical prismatic element. The direction of the decoupled rays from the light guide and the efficiency of the reflection on each prism are given by the acceptance angle β (rad) between the input surface of the prism and the wall of the light guide. The acceptance angle (β) of an individual prism changes with its distance from the source [20].

The angles of prismatic surfaces (β), as shown in Figure 4.24b, can be arranged to divert the light rays from the LED source toward the desired direction. Therefore, such prismatic surfaces allow to control the light distribution.

The light rays coming from a source are kept in the transparent light guide by TIR, which is defined as 100% reflection of the rays on a surface between an optical medium with a high and low index of refraction. The refractive index of the surrounding material, which is air (n_2) for automotive applications, is

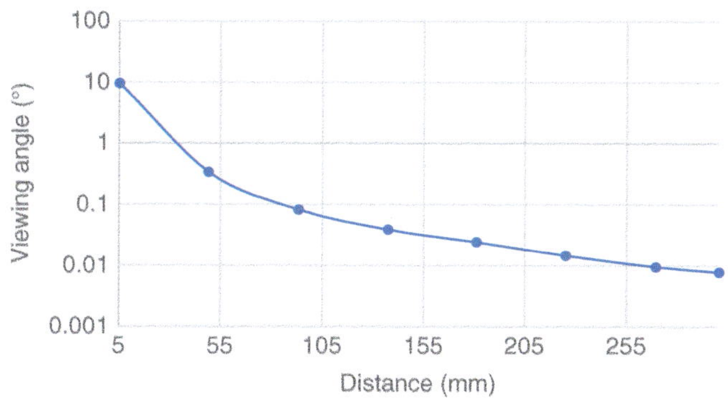

Figure 4.23 Dependence of viewing angle on the distance along with the light guide. Source: Wördenweber et al. [7]/Springer Nature.

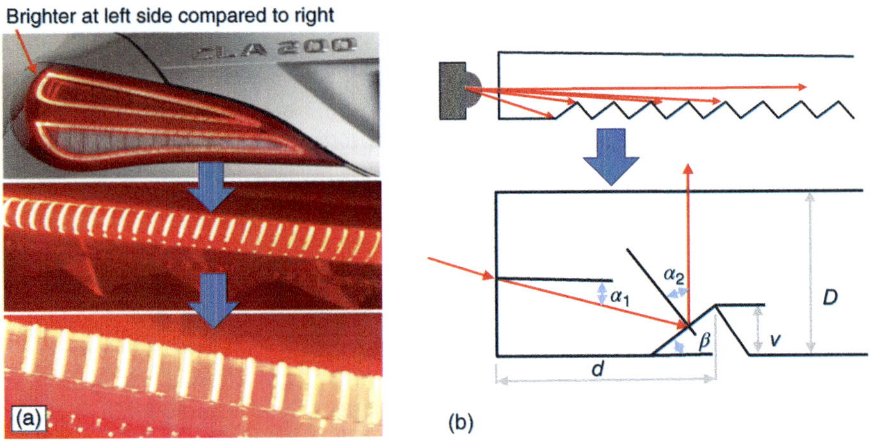

Figure 4.24 Typical prismatic element light guide showing the incident light and prismatic surface angle [7, 20]. (a) Light guide illuminated with LED module in the CLA200 Mercedes Benz. (b) LED ray in prismatic structure light guide. Showing incident angle α and prismatic surface angle β.

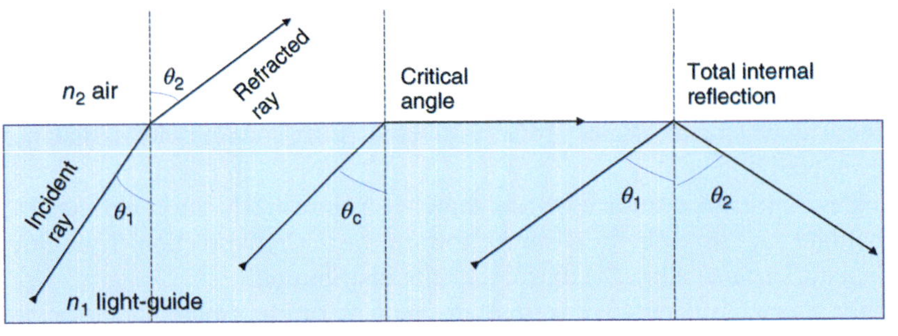

Figure 4.25 Critical angle and total internal reflection.

always lower than the refractive index of the light guide material (n_1). TIR occurs if the incident angle (θ_1) is bigger than the critical angle θ_c, as shown in Figure 4.25.

This critical angle (θ_c) is derived using Snell's law. When the angle of the incidence is equal to the critical angle, the angle of refraction (θ_2) will be equal to 90°. Using Snail's law as shown below, we can derive the critical angle, θ_c:

$$n_1 \sin \theta_c = n_2 \sin \theta_2, \quad \text{Snail's Law.}$$

When $\theta_2 = 90°$, n_2 = index of refraction of air = 1, n_1 is the light guide refractive index.

$$\theta_c = \sin^{-1} [1/n_1], \tag{4.1}$$

TIR occurs if the incident angle θ_1, is bigger than the critical angle (θ_c). Using Eq. (4.1), the critical angle can be estimated. For example, for clear poly(methyl methacrylate) (PMMA) with a refractive index of $n_1 = 1.493$, the value of the critical angle is 42°.

In order to have 100% TIR, the angle β of the prismatic element of the light guide must be greater than θ_c. On the other hand, the direction of the decoupled rays and the efficiency of the reflection on each prism are also influenced by the β angle (rad), which is the angle between the input surface of the prism and the wall of the light guide, as illustrated in Figure 4.24b [20]. Angle β is estimated using Eq. (4.2):

$$\beta = \frac{\frac{\pi}{2} - \text{arctangent}\left[\frac{D/2-v}{d}\right]}{2} \tag{4.2}$$

where D is the diameter of the light guide, v is the depth of the prism, and d is the distance of the prism from the entrance surface of the light guide.

Equation (4.2) provides good results for direct light from a light source, especially at the beginning of the light guide, as illustrated in Figure 4.24b. On the other hand, in the center and end of the light guide, as illustrated in Figure 4.26, it is more beneficial to use an equation that considers one reflection on the sidewall of the light guide [20]. This equation for estimating the angle β_i is shown in Eq. (4.3):

$$\beta_i = \frac{\frac{\pi}{2} - \text{arctangent}\left[\frac{D/2-v+D}{d}\right]}{2} \tag{4.3}$$

The efficiency of the reflection on each prism depends on the incidence angle β_i, given by the Fresnel coefficient [21]. The efficiency continuously increases with the distance from the entrance surface of the light guide to 100% of the TIR, as illustrated in Figure 4.27.

Although the first few elements receive much more energy directly from the light source, the efficiency of the reflection is lower than in the middle of the light guide, where the prism elements are calculated assuming one reflection along the length of the light guide. The light's uniformity can be further improved by shaping the prismatic structure. This is possible by adjusting the parameters (βi, v, d, and D) of a light guide to find the optimum values for the light distribution [7].

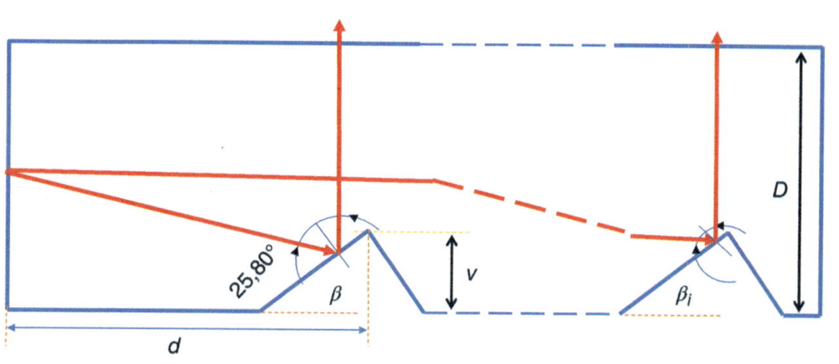

Figure 4.26 Different incident angles at the beginning and end of the light guide. Source: Adapted from Wördenweber et al. [7] and Güney et al. [20].

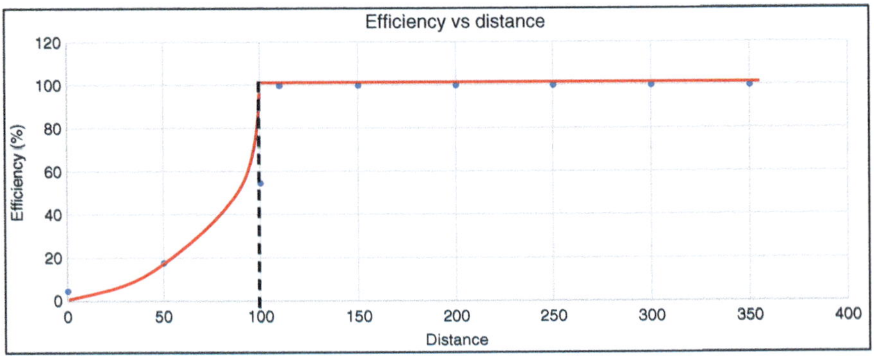

Figure 4.27 Efficiency of light reflection by the prismatic structure. Source: Wördenweber et al. [7]/Springer Nature.

4.4 Interior Lighting

Automotive interior lighting has captivated many car buyers. It is one of the key attractions for car buyers. Some even say that their car is their mobile living room. Having said that, automotive interior designers have gone to great extent to design the ambient condition conducive enough to fulfill such expectations. Having good interior lighting certainly helps to attain such expectations, besides other comforts like sound system, comfortable leather seating, noise cancellation, and good suspensions. Figure 4.28a,b shows examples of interior lighting of luxury cars. These interior lighting consists of light guides with edge-lit by Red Green Blue (RGB) LED.

Automotive interior lighting illuminates the cabin when people enter the vehicle or dismount and thus aids passengers to fasten seat belts and turn on the ignition. Dashboard lights help display important vehicle parameters and warning signs.

LEDs are the key light sources in the field of interior lighting. Already a long-established fixture of the instrument panel, these lights can now also be used for other interior lighting, like reading lamps. Thanks to the versatility of white and colored LEDs – which also offer low power consumption compared to filament bulbs. This opens a whole range of decorative design possibilities for the cockpit in addition to the functional lighting aspect.

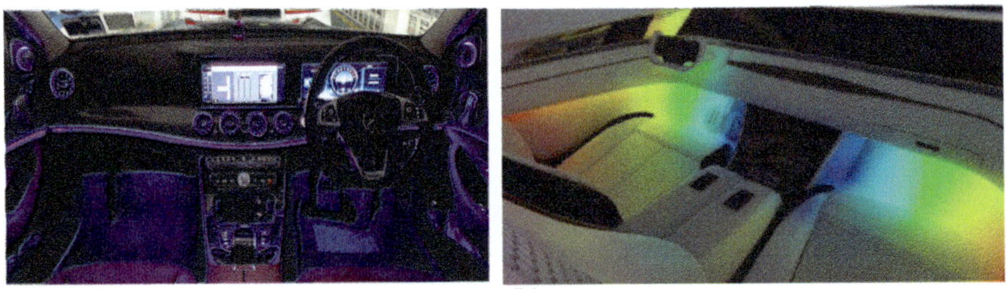

(a) (b)

Figure 4.28 Interior lighting – automotive ambient lighting. Source: Courtesy of ams OSRAM GmbH.

As in architectural lighting, we use the term "ambient light" to describe a low light level that helps us perceive the surrounding space. It also serves a decorative function, and skillful ambient light design can emphasize specific interior design features or achieve surprising effects when these lights are used unexpectedly behind translucent material.

The automotive interior lighting has followed the general trends in lighting and will therefore evolve toward many RGB LEDs per car in future design. Creating mood at day and night, pixelated sign-like information, and theater-like effects like welcome and goodbye messages are examples of outstanding user experience.

4.5 Summary

Automotive lighting is one of the main drivers of the LED industry and technology. Over the years many innovations in LED have shaped the automotive lighting technology to an unparallel level of any other light source achieved, and yet a lot more innovation is still coming out of the automotive lighting industry. As consumers, we are fortunate to see the amazing innovation coming from the automotive lighting makers as well as from the LED light source makers. These combined innovations make the future of automotive lighting very bright.

References

1. Schubert, F.E. (2006). *Light Emitting Diode*, vol. 2. New York, USA: Cambridge University Press.
2. OSRAM (2020). *OSRAM product data sheet – OSLON®SX – LA CN5M*, V. 1.6, Editor. OSRAM Opto Semiconductors. Bayan Lepas, Malaysia.
3. OSRAM, Wilm, A., and Chew, I. (2019). *Light quality - White Light Parameters*. OSRAM.
4. OSRAM (2021). *LED ENGIN LuxiGen*. OSRAM.
5. OSRAM (2020). *OSRAM LED training material*. Bayan Lepas, Malaysia.
6. Fryc, I., Czyżewski, D., Fan, J., and Gălăţanu, C.D. (2021). *The drive towards optimization of road lighting energy consumption based on mesopic vision—A suburban street case study*. *Energies* 14 (4): 1175.
7. Wördenweber, B., Wallaschek, J., Boyce, P., and Hoffman, D.D. (2007). *Automotive Lighting and Human Vision*, vol. 1. Springer.
8. Schmidt, C., Kalze, F.J., and Eichhorn, K. (2007). Adaptive front-lighting system in LED technology–initial steps and the future. *SAE Transactions* 332–342.
9. Adhav, P.V. and Shaikh, S. (2014). Adaptive front lighting system using CCD. *IOSR Journal of Electronics and Communication Engineering* 9 (5): 20–25.
10. Moore, D.W. (1998). *Headlamp history and harmonization*. University of Michigan, Ann Arbor, Transportation Research Institute.
11. Kobayashi, S. (1998). Intelligent lighting systems: their history, function, & general direction of development. *SAE Transactions* 107: 1798–1806.

12 Ley, P.-P., Held, M.P., and Lachmayer, R. (2018). *Analysis of LED arrangement in an array with respect to lens geometry*. In: *Light-Emitting Diodes: Materials, Devices, and Applications for Solid State Lighting XXII* (ed. J.K. Kim, M.R. Krames, M. Strassburg, and L.-W. Tu), 1055405–1055405-8. International Society for Optics and Photonics.

13 Group, H (2014). HELLA matrix LED system. [cited 8 2], 78. https://youtu.be/xYSix5r38qY (accessed 8 April 2021).

14 Taki, T. and Strassburg, M. (2020). Visible LEDs: more than efficient light. *ECS Journal of Solid State Science and Technology* 9 (1): 015017.

15 Anything, R. (2020). *All new 2021 Mercedes S class – digital light*.

16 Morkos, B., Shankar, P., Teegavarapu, S. et al. (2009). Conceptual development of automotive forward lighting system using white light emitting diodes. *SAE International Journal of Passenger Cars-Electronic and Electrical Systems* 2 (2009-01-0593): 201–211.

17 Rumar, K. (2003). *Functional requirements for daytime running lights*. University of Michigan, Ann Arbor, Transportation Research Institute.

18 Le, H.-T., Le, L.-T., Liao, H.-Y. et al. (2020). Design of low-glared LED rear light of automotive for EU ECE regulation by use of optimized micro-prisms array. *Crystals* 10 (2): 63.

19 Stewart, J.W. (1998). HP SnapLED: LED assemblies for automotive signal lighting. *Hewlett-Packard Journal* 50 (1): 1.

20 Güney, E., Alper, M., and Hacıismailoğlu, M. (2020). Optical design of light guide prisms with surface roughness for automotive tail lights. *Proceedings of the Institution of Mechanical Engineers, Part D: Journal of Automobile Engineering* 0954407020907209.

21 Born, M. and Wolf, E. (2013). *Principles of Optics: Electromagnetic Theory of Propagation, Interference and Diffraction of Light*. Elsevier.

5

LED Application For Consumer Industry

5.1 Consumer Indoor Lighting

The TV applications were the main driver in display mass market in late 1980s; however, this was changed by the game changers such as emergence of computers, laptops, and mobile telephones in late 2000s. As the consumer demand for high-efficiency LEDs in terms of power savings and better lighting quality has propelled the unparalleled LED technology development over the past decades. As a result, LED holds large market within electronic devices and generates income for countries economically.

Longer lifespan and better reliability boost its wide applications in indoor and outdoor surroundings. Moreover, LEDs consume less power while delivering almost 50,000 hours of light compared to their predecessors, the incandescent lights. Lower power consumption also means lower heat generation, which complies with most energy-saving policies around the world. This spurred aggressive demand for LEDs, and eventually increase in supply decreased the prices. Into the bargain, the new strict energy policies across European Union, China, and United States favor LED-based products for future. Government rebates on power consumption further propel all manufacturing companies to favor energy-saving lighting in their offices and fabrication floors.

By the name, indoor lighting would mean any form of artificial lighting used without any interference from natural weather such as rain, humidity, dryness, and/or man-made weather such as fog from traffic, mining, and many others. Historically, the main usage of indoor lighting was in manufacturing floors during industrial revolution. This has been more intense since industrial boom in automobile manufacturing, telecommunications, tourism, general offices, and also in educational areas. With such different areas where this lighting has been used, it require different specifications and requirements to fulfill them. This can be categorized as:

1- Manufacturing floors, such as line assembly and controlled device fabrication, need specific lights.
2- Health care and medical treatments
3- General office lighting

LED Packaging Technologies: Design, Manufacture, and Applications, First Edition.
Luruthudass Annaniah, Mohamed Salleh M. Saheed, and Rajan Jose.

4- Sensors in food industries for chemical composition detections.
5- Safety and security such as face recognition and advanced iris recognition.

The consumer indoor lighting in manufacturing and general office was discussed extensively in many reviews. However, the current advancement of LED usage in biosensors, health, and security is still expanding. We will focus on this new LED area of growth to illuminate the readers on the importance of LED devices in health care and security usage.

5.2 Health Care and Medical Treatments

In the past, lasers were extensively used for health care and medical treatments; however, these lasers pose possible hazards to human cells [1]. Moreover, the lasers are rather expensive, and their intricate controllability increases their overall production and limits their usage. Here, the LEDs have much to offer in imaging and photobiomodulation (PBM) since they can be mass produced, have less complex integration, and are safer compared to lasers. According to the dictionary, the PBM brings the meaning of nonthermal process by placing light source near the target, where this electromagnetic wave penetrates into cells or tissues and interacts with the chromophores, resulting in various biological levels. To achieve the desired outcome from PBM technique, the LEDs need to fulfill two main criteria, which are light coherence and energy densities [1]. In the past, unlike lasers, LEDs had rather wide light coherence. With current LED chip advancement, the LEDs are able to reach significant light coherence, such as 200 lm/W cool white LEDs. To make sure the PBM method does not work thermally, the energy densities are defined to be between 5 mW/cm^2 and 5 W/cm^2, so LED is easily able to achieve this with current commercially available. Additionally, applying LEDs in arrays, increases the beam area, which is suitable for large-area treatment.

However, the LEDs outputs need to be well controlled to achieve the targeted outcome. On the other hand, the overdose of light exposure, may lead to adverse complications such as cell destruction or toxicity [2, 3].

Here, we will discuss the LED-based therapies for diabetic foot ulcers and skin reconstruction. The signal activation for protein synthesis, cell regeneration, and repair can be achieved via PBM by using certain light wavelengths. By understanding the specific activation energy needed for cells responsible for skin reconstruction, the desired LED can be chosen for the application. In the past, skin exposed to 660 nm LED wavelength showed reduction in the enzymes responsible for peptide bond destruction and increased cellular response at cell surface, which led to skin reconstruction [4]. Other studies also reported skin reconstruction by using LEDs with 630–660 nm wavelengths and adjusted energy density per area of 4 J/cm^2 [5].

Similarly, to speed up the wound closure, light has extensively been used. The increased activity of lymphocytes and cell proliferation were seen to accelerate wound closure. The key factor here is that increased fibroblast cell growth can be achieved via light exposure from visible to near infrared wavelengths [5]. The effectiveness of LED light therapies has been reported in the multiple combinations of three different light wavelengths, such as 670, 720, and 880 nm [6]. Recent study also suggests that light irradiation of 5 J/cm^2 at 660 nm accelerates wound closure by enabling angiogenesis and suppressing volatile oxygen production within cells [7].

In recent COVID-19 pandemic, the use of pulse oximetry has become an important device in every household, not limiting to only health care facilities as in the past. A sensor consisting of photodetector is placed near location of LED illumination to detect the non-absorbed light. Such methodology is known as photoplethymysography (PPG), which is a nondestructive technique made up of photodiode-LED arrangements [8]. The simple working principle of LED-photodiode arrangement is that the emitted light through human body is rather absorbed, reflected, or scattered. These light modes were later detected by using photodiode, which acts as the probe or sensor placed on human body. The absorbed, reflected, or scattered light leads to a change in light intensity; these distinctions are measured, filtered, and amplified, and a useful voltage signal is produced. Pulse oximetry is a non-cell destructive measurement device to detect the oxygen saturation level (SpO_2). By detecting the dissolved oxygen level in blood hemoglobin and deoxyhemoglobin, the oxygen saturation, SpO_2 was defined [9]. These two different hemoglobins have different wavelength absorption when light is illuminated. Namely, at 660 nm wavelength, the deoxygenated hemoglobin has higher absorption in the red-light spectrum, while at 940 nm, the oxygenated hemoglobin shows high light absorption, as shown in Figure 5.1.

Similarly, in case of pulse oximeter, the photodetector is used to measure the variation of non-absorbed wavelength from the LEDs. These excitation signals were

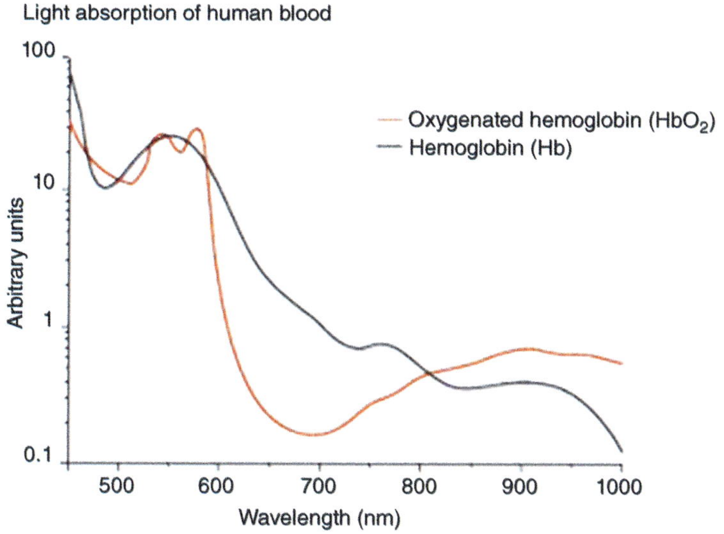

Figure 5.1 Light wavelength absorption in oxygenated and deoxygenated hemoglobin (human blood) [10]. Source: Courtesy of ams OSRAM GmbH.

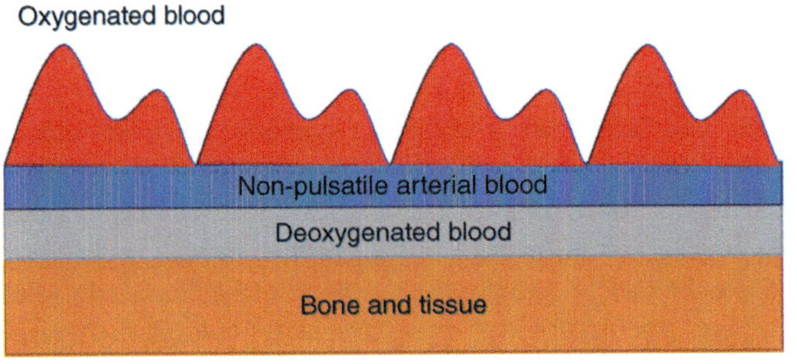

Figure 5.2 The AC and DC components at wavelength exposure.

amplified into two main electrical signals: AC and DC components. The SpO_2 level was the ratio of deoxygenated AC/DC at 660 nm to oxygenated AC/DC at 940 nm as in equation 1.

$$SpO_2 = \frac{\left(\frac{AC}{DC}\right)660}{\left(\frac{AC}{DC}\right)940} \tag{5.1}$$

The AC and DC components are depicted in Figure 5.2. The AC is assigned to pulsatile arterial blood, while DC is assigned to non-pulsatile arterial blood, deoxygenated blood, bone, and tissue.

A system setup with LEDs and photodetector was the preferred design due to its design freedom and simplicity. Here the LED design plays important role to have maximum light output, resulting in accurate measurement of SpO_2. Two ways to minimize the LED light output are by using reflective package and chip with top-side light emit, as discussed in detail in previous Chapters 2 and 3.

5.3 Safety and Security

5.3.1 Led in Iris Recognition System

As technology keeps on evolving and different types of electronic devices are being invented, questions regarding the credibility of their security systems may arise. A good security system will help an individual, a corporation, or a government to secure all the information and assets that are valuable to them. For example, an individual needs to secure his communication devices, such as handphones and laptops, which may contain important and confidential information. The most common security systems are pin numbers, passwords, fingerprints, facial recognition, and iris recognition. Of all this, iris recognition system has an advantage in terms of its security.

Iris recognition seems to be the best because of its unique features, such as being fast, simple, and accurate. The uniqueness of iris helps to identify and verify the identity of a person much more accurately and faster than other biometric systems. Iris is located in between the lens and cornea of our eyes. Iris's task is to

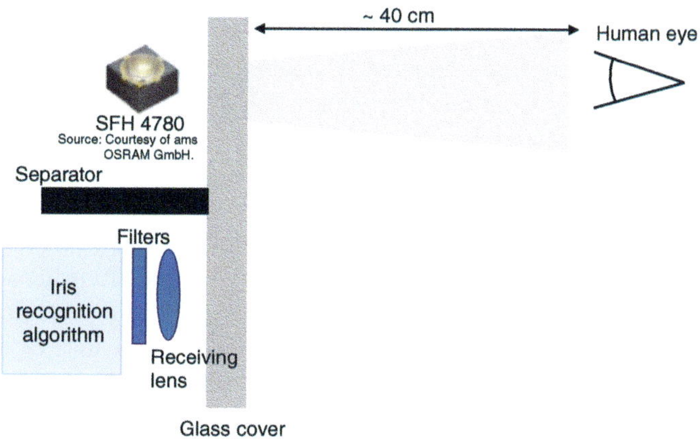

Figure 5.3 System hardware used in for the Iris recognition.

control the amount of light that penetrates the eyes by opening and closing the pupil.

Iris recognition system contains only four main parts, which are illumination module, camera (includes lens and image sensor), optical bandpass filter, and a software algorithm, as shown in Figure 5.3. All of these components are mounted into a glass cover, resulting in a relatively smaller footprint sensor compared to the fingerprint sensor. The essential part that determines the quality of an iris scanner is the optical system, which uses a suitable emitter.

Depending on the differences in human eye colors, different light wavelengths are required. Commonly, there are two wavelengths that are being used, which are infrared and visible wavelengths. For humans with dark and brown eyes, visible wavelength is not suitable for iris scanning as it has a poor characteristic for extracting the iris details. Alternatively, for people with blue and green eye colors, both infrared and visible wavelengths are deemed to be the perfect wavelengths for iris scanning [11]. However, visible wavelength is able to expose the unique characteristics of iris more precisely compared to infrared wavelength. Here, the SFH 4780S with 810 nm emitter offers the best solution for all eye colors.

To synchronize these differences, an integrated wavelength that is able to detect all types of eye colors around the world is needed. Thus, a single high-power 810 nm is required. This module can maximize the signal-to-noise ratio whenever the iris needs to be scanned in the bright place, in which case the filter in the camera has to use narrow optical bandpass filter. This module gives an excellent result in revealing a unique structure of a human iris. An example of an 810 nm product that is available on the market is OSLUX SFH 4780S, a product of OSRAM Opto semiconductors [11].

The high irradiance amount in SFH 4780S gives a perfect picture despite the lighting conditions in particular place. For example, on a sunny day, this scanner is able to detect the iris of a human efficiently. Moreover, with a good thermal management, this module can be utilized efficiently since it has lower thermal resistance (Rth) range.

5.3.2 LED in Food Processing

One of the biggest challenges in food industry is maintaining its food safety standards. As foods are essential part of our lives, the demand for them is always increasing. Thus, its safety will always be a concern for both the consumer and the food producer. Generally, a good food safety level can be achieved when the foods (could be solid food or liquid food) are processed in hygienic manner and distributed to consumers in a proper way. Hygienic in this context means that the foods are handled with proper care from the beginning of the process. To achieve a proper standard of food safety, every part of the food processing chain plays an important role. The end product should be delivered to the consumer in a safe and hygienic condition. Failure in keeping the food hygiene standard may cause significant impact to the producer and the consumer. The food producer might be convicted by authorities if they breach the food preparation acts. On the other side, if the consumer consumed foods that did not comply with the standard, for example, contaminated foods, they could suffer from food poisoning and may suffer from foodborne illnesses, such as Norovirus and Salmonella.

Before the emergence of LED technologies, artificial light treatment was widely used in food industry to sanitize and disinfect water and foods. Examples of artificial light treatments that were used back then were mercury vapor lamps and xenon lamps. These lamps were used to kill the bacteria, yeast, viruses, and fungi that may exist in foods or in water. Even though these lamps were able to inactivate the above microorganisms, they left some mercury residues in the food and water, which could be harmful to consumers if consumed excessively and continuously. This lamp also does not have longer life span and also consumes much more electricity if compared to LED equipment when treating foods and water [12] (Figure 5.4).

LED equipment, on the other hand, is an excellent tool in terms of its safety and its life span. By using monochromatic light and different wavelengths, it can inactivate microorganisms such as bacteria, yeast, viruses, and fungi without leaving any unwanted substances. Since the size of LED tools is compact and flexible, making it very convenient to use in any tool, whether to kill a certain pathogen i.e. E-Coli, or to deactivate all the microorganisms that live in the foods as a whole treatment [14].

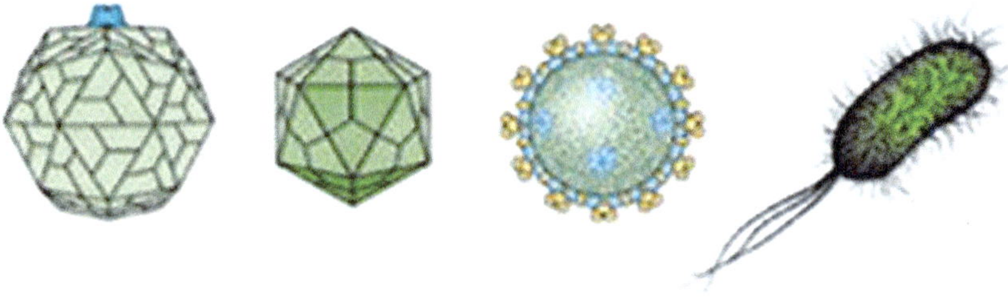

Figure 5.4 Some types of microorganisms, which absorb specific wavelengths and are unable to multiply further. Source: Adapted from Florent et al. [13].

5.3.3 Treatment in Solid and Liquid Foods

Fruits, fish, poultry, rice, and grains are some examples of solid foods. These foods are processed and packed into tins, glasses, or heat-resistant plastics. The crucial part of this cycle is the sanitization process. The process should be done accordingly in order to prevent the solid foods from being contaminated with any microorganisms, which may spoil the products. Microorganisms that exist in foods can survive for a long period in their static mode and immediately activate when they are exposed to certain conditions [15]. Foods like raw meat and poultry contain high numbers of bacteria, such as *E. coli* and *Salmonella*. Therefore, it should be cooked at an appropriate temperature in order to keep its nutritional value and also to kill the microorganisms. In order to achieve a maximum impact of the LED equipment in decontaminating the foods, several aspects should be taken into consideration. Some of them are the nature of the food, level of water activity (aw), and the structure of the food. Depending on those aspects, repetitive or pulsed LED treatment can be chosen.

Liquid foods such as fruit juices, milk, soups, and coffee are very susceptible to contamination by the microorganism. This is mainly because of the water level (aw) of the food and also because of its high carbohydrate content, especially sugar and fiber [15]. To increase the liquid foods' lifespan, artificial preservatives are added, besides food coloring to enhance the visual effects of the foods. However, as people are beginning to be aware of the effects of certain artificial preservatives on their health if consumed excessively and continuously, the demand for the preservative-free liquid foods is increasing. Thus comes the need for LED base treatment. It is a known fact that LED treatments are efficient in preserving foods without leaving significant odor or residues in the liquids.

One of the most used LED tools in food sterilization is UV-C. Whenever the UV-C light hits the food, its DNA and RNA will captivate the radiation, thus encouraging the creation of thymine dimers as shown in Figure 5.5. This formation of thick thymine dimers will stop the duplication of the microorganism and deactivate the bacteria, fungi, or viruses, resulting in safe and clean food products.

Figure 5.5 The natural UV light used to damage the RNA construction, can be artificially created with UV LEDs. Source: Adapted from Saha and Chen [16].

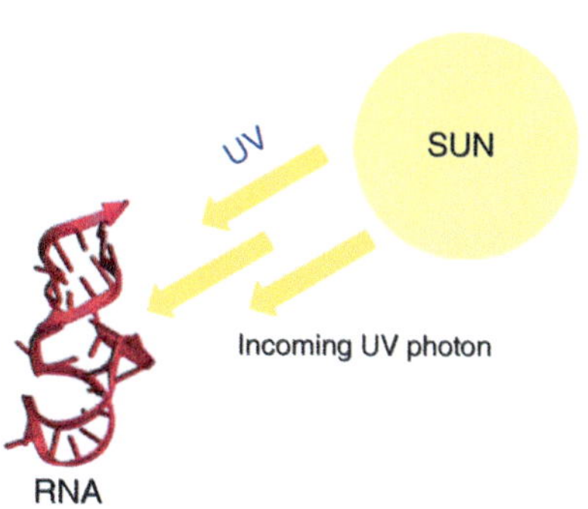

5.3.4 Water Treatment

Water is an important element for human beings, animals, and plants. Particularly for human beings, consuming clean water is essential to provide hydration to the body. However, not everyone has the privilege to consume clean water since it is expensive to treat wastewater. Improper water treatment may cause water-borne diseases such as cholera, hepatitis A, and diarrhea. Chemically treated water using chlorine produces chlorine disinfection by-products, such as trihalomethanes and haloacetic acids, which could increase health risks to consumers with prolonged usage [17].

LED water treatment using UV-C radiation is very affordable because of its low voltage level and also because it is very efficient in deactivating the pathogens without leaving any significant residue or odor. By using appropriate wavelength, chemical-free water can be obtained and consumed. The more health-conscious consumer is now opting for safer way to disinfect any food and drinks compared to conventional chemically treated products. As we have seen before, the LEDs were mainly focused on providing lighting element to the consumer; however, as the technology matures, the use of LEDs in daily life also broadens.

References

1 Dong, J. and Xiong, D. (2017). Applications of light emitting diodes in health care. *Annals of Biomedical Engineering* 45 (11): 2509–2523.

2 Sharma, S.K., Kharkwal, G.B., Sajo, M. et al. (2011). Dose response effects of 810 nm laser light on mouse primary cortical neurons. *Lasers in Surgery and Medicine* 43 (8): 851–859.

3 Hennessy, M. and Hamblin, M.R. (2017). Photobiomodulation and the brain: a new paradigm. *Journal of Optics* 19 (1): 013003.

4 Wang, C., Tsai, S., Yu, M. et al. (2015). Light-emitting diode irradiation promotes donor site wound healing of the free gingival graft. *Journal of Periodontology* 86 (5): 674–681.

5 Barolet, D. (2008). Light-emitting diodes (LEDs) in dermatology. *Seminars in Cutaneous Medicine and Surgery* 27 (4): 227–238.

6 Whelan, H.T., Smits, R.L., Buchman, E.V. et al. (2001). Effect of NASA light-emitting diode irradiation on wound healing. *Journal of Clinical Laser Medicine & Surgery* 19 (6): 305–314.

7 Beckmann, K.H., Meyer-Hamme, G., and Schröder, S. (2014). Low level laser therapy for the treatment of diabetic foot ulcers: a critical survey. *Evidence-based Complementary and Alternative Medicine* 2014: 1–9.

8 Stojanovic, R. and Karadaglic, D. (2013). Design of an oximeter based on LED-LED configuration and FPGA technology. *Sensors (Switzerland)* 13 (1): 574–586.

9 Freescale Semiconductor, Inc. (2011). Pulse Oximeter - Fundamentals and Design.

10 Stefanie, R. and Florian, L. (2021) Health monitoring Valid for: BIOFY ® / TopLED ® D5140 / Chip LED ® Sensors / Firefly ® E1608 / Firefly ® E2218 / PointLED ®.

11 Hanna, B. and Alfons, S. (2018) IR OSLUX ®-SFH 4780S in iris recognition applications Valid for: OSLUX ® SFH 4780S.

12 Vilhunen, S., Särkkä, H., and Sillanpää, M. (2009). Ultraviolet light-emitting diodes in water disinfection. *Environmental Science and Pollution Research* 16 (4): 439–442.

13 Florent, P., Cauchie, H., Herold, M., and Ogorzaly, L. (2022). Bacteriophages pass through candle-shaped porous ceramic filters: application for the collection of viruses in soil water. *MicrobiologyOpen* 11 (5): e1314.

14 Chatterley, C. and Linden, K. (2010). Demonstration and evaluation of germicidal UV-LEDs for point-of-use water disinfection. *Journal of Water and Health* 8 (3): 479–486.

15 Prasad, A., Du, L., Zubair, M. et al. (2020). Applications of light-emitting diodes (LEDs) in food processing and water treatment. *Food Engineering Reviews* 12 (3): 268–289.

16 Saha, R. and Chen, I.A. (2019). Effect of UV radiation on fluorescent RNA Aptamers' functional and templating ability. *ChemBioChem* 20 (20): 2609–2617.

17 Botlagunta, M., Singh Bondili, J., Js, B., and Mathi, P. (2015). Bioelectronics view project role of integrins in cellular transformation View project WATER CHLORINATION AND ITS RELEVANCE TO HUMAN HEALTH. *Article in Asian Journal of Pharmaceutical and Clinical Research* 8.

6

LED Application for General Lighting

Ever since mankind mastered the creation and use of artificial light, their civilization has evolved in great leaps and bounds. Artificial light has given humans an edge over other species in this world and has evolved over time. At the end of twentieth century, an invention by Shuji Nakamura in creating artificial light changed the artificial lighting industry, and along with it, many new inventions followed. Nakamura invented the high-brightness blue light-emitting diode (LED) using gallium nitrate material system in the LED chip [1], for which he received the Nobel Prize in Physics in 2014. Using blue light and phosphor converter material, the blue light was converted to white light. This invention has changed the whole lighting industry [2].

The LED is known for its long lifetime, robustness, and energy efficiency; LEDs have already replaced many conventional incandescent and fluorescent technologies in many lighting applications. Especially in general lighting applications.

The LED application for general lighting can be divided into two market categories, as shown in Figure 6.1. One is retrofit lighting, which interfaces with existing socket. Retrofit lighting market covers the residential, office, shops, and hospitality and covers indoor or decorative lighting as well as artistic and architectural lighting. Second is LEDfit lighting, which uses luminaires/modules with built-in LEDs. LEDfit lighting market covers shop, office, residential, and outdoor, like street lighting, stadium lighting, artistic and architectural lighting, and horticulture lighting. In next subchapters, some of these lightings will be further elaborated.

6.1 RETROFIT Lighting

In this subchapter, we briefly explain the retrofit lamp construction with pictures to illustrate the amazing adaptation that an LED can provide for the lighting industry. Here, the retrofit lamp design and the stages of energy flow will be explained in detail. Beside the retrofit lamp, this subchapter also briefly explains the architectural lighting application in the hospitality industry.

LED Packaging Technologies: Design, Manufacture, and Applications, First Edition.
Luruthudass Annaniah, Mohamed Salleh M. Saheed, and Rajan Jose.

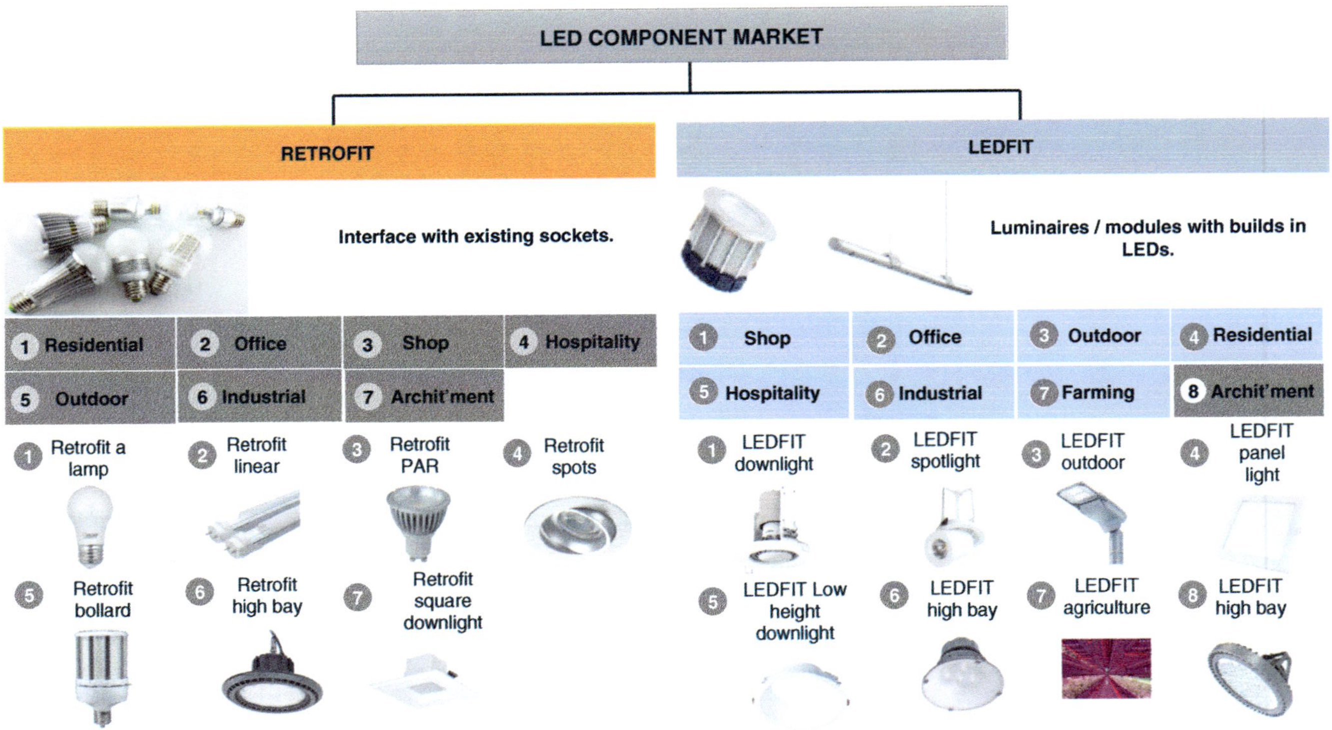

Figure 6.1 LED general lighting market. Source: Courtesy of ams OSRAM GmbH.

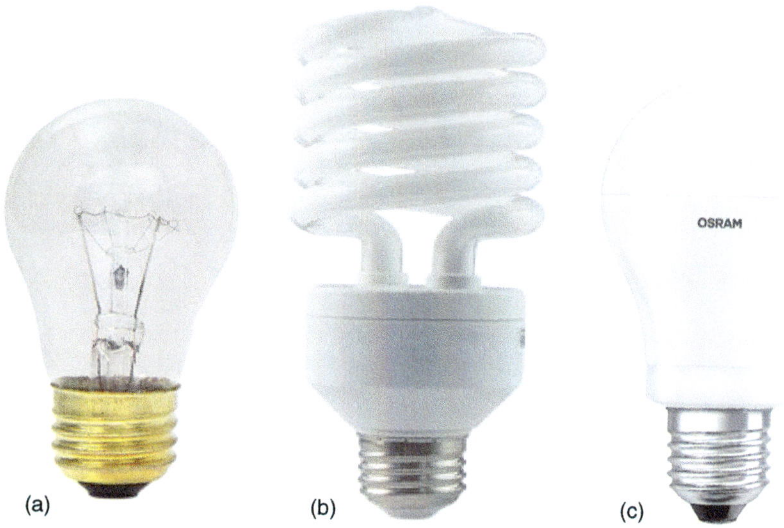

Figure 6.2 Conventional lamp – (a) incandescent lamp, (b) CFL lamp, and (c) equivalent LED lamp. Source: Courtesy of ams OSRAM GmbH.

6.1.1 RETROFIT Lamp

LED has been widely applied in interior lighting applications. For almost all the lamps on conventional technologies, an LED equivalent has been developed and marketed as their LED base lamp. One good example is if we look into the retrofit lamp applications as shown in Figure 6.2: (a) conventional lamp, (b) compact fluorescent lamp (CFL), and (c) an equivalent LED lamp. The retrofit LED lamps are designed to be the replacement for conventional lamps and have the advantage that they can be applied to existing luminaires. This allows users to upgrade their lighting system to LED-based technology without having to replace the whole luminaire.

The requirement that a retrofit lamp has to replace an existing conventional light source puts many limitations on the design of the lamp. The electrical, mechanical, and optical interfaces are adjusted to fit the LED requirements to fit the retrofit lamp, i.e. it must fit the lamp socket. It has to fit a certain shape and provide a certain amount of light that is comparable to light distributions and light quality. The voltage supplied by conventional lamps is usually higher than that required for LED lamps. Hence, the LED lamp has to be equipped with integrated electronic circuit to convert the supplied voltage of 230 V to a voltage and current that an LED light source can handle. All these aspects need to be considered when designing a retrofit LED lamp.

Among the retrofit light bulb products, the most common lamp in use at home is the E27, as shown in Figure 6.3. The E27 lamp is also known as Edison bulb with big screw socket (27 mm). The E27 refers to a socket, the fastening that you screw into the light appliance.

The design of these LED lamps is unlike traditional incandescent lamps that use tungsten filaments with high melting temperatures without the need for the

Figure 6.3 E27 retrofit LED lamp. Source: Courtesy of ams OSRAM GmbH.

heatsink. LED is a semiconductor device that has low melting temperature and therefore requires a heatsink to keep the junction temperature below a certain temperature limit that is defined by the LED junction temperature (typically 125 °C). Besides the thermal issue, LEDs have a thermal droop characteristic because its luminous efficacy decreases as the junction temperature increases [3]. Several attempts have been proposed to consider the interaction of photometric, electric, and thermal aspects of LED devices. Such thermal-dependent photometric behaviors have been characterized mathematically for LED systems by photo–electro–thermal theory. The characterization result shows the luminous output of the compact LED lamp has a strong dependence on ambient temperature [4]. Hence, we need to understand the energy flow in the LED lamp design, to design a high-efficacy LED lamp.

Commonly, an LED lamp system consists of two spectra: the first is the blue light spectrum generated from the blue LED that comes from InGaN material system at typical 450 nm wavelength, and the second is the light spectrum of Stokes-shifted wavelengths emitted from a yellow phosphor converter. In the Stoke-shift process, the phosphor absorbs blue light and converts it to white light [5]. There is a loss of heat energy generated during this light conversion process. This is commonly known as the Stokes-shift loss.

The Figure 6.4a shows the detailed breakdown of an LED lamp. The chip is coated with yellow phosphor and is soldered on the PCB. The PCB is attached to heatsink with thermally conductive glue to dissipate heat that was generated by the LED and phosphor to ambient. Figure 6.4b illustrates a cross-section view of an LED lamp that shows phosphor coated on an LED on a substrate that has heatsinks attached at the bottom and socket with ballast inside. For simplicity, the phosphor-based LEDs can be perceived as having two power processing substages–one being the blue LED chip generating blue light and the other being the phosphor layer performing the light conversion process.

The energy flow in an LED lamp can be described in stages. In Figure 6.5, these energy flow stages are described in detail [6].

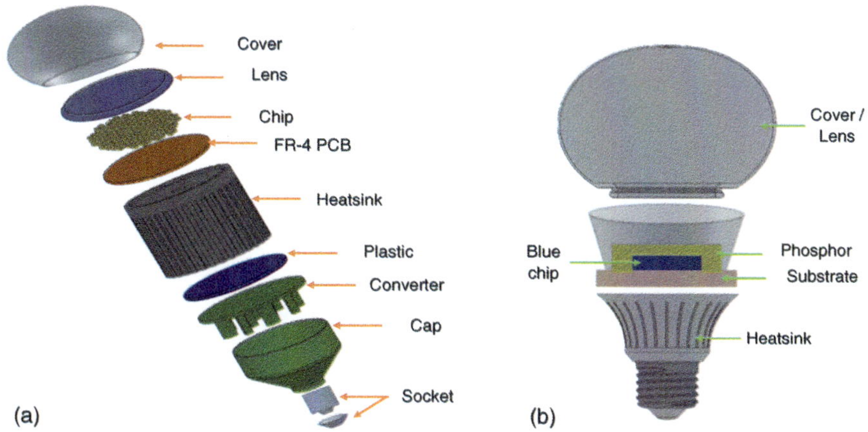

Figure 6.4 Construction of an LED-based lamp. (a) Detailed design breakdown of LED lamp. (b) Showing the phosphor-coated LED chip in side and LED lamp.

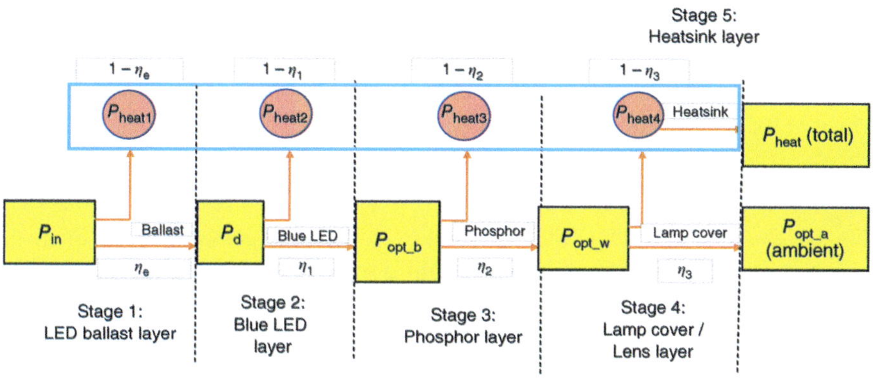

Figure 6.5 Stages of energy flow in an LED-based lamp system.

At Stage 1, the input power P_{in} is processed by the LED ballast with an efficiency of η_e. An LED ballast is usually a single-stage switched-type ballast employed in a compact LED system design due to its high-frequency operation. Usually, the ballast comprises an input diode bridge and a cascaded AC/DC converter. The power delivered to the LED chips from the ballast can be generalized as $P_{in,}$ which is the input power. The diode power P_d can be described by Eq. (6.1).

$$P_d = P_{in}{}^{*}\, \eta_e \tag{6.1}$$

where η_e is the LED ballast efficiency.

At Stage 2, the blue LED chip converts electrical energy into light energy by emitting blue light at an efficiency of η_1 (η_1 is known as the extraction efficiency, to be detailed later). The optical power of the blue LED, $P_{opt\text{-}b}$, is given by Eq. (6.2):

$$P_{opt\text{-}b} = P_d{}^{*}\, \eta_1 \tag{6.2}$$

While the rest of the input power, $P_d(1-\eta_1)$, is converted into heat. The heat generation is related to several power loss mechanisms, such as the leakage current power lost due to tunneling of electrons at the epitaxy interface, auger recombination, and non-radiative recombination.

Additionally, any photons generated by radiative recombination inside the LED chip may be emitted as external light or trapped within the LED chip (caused by the total internal reflection phenomenon of the semiconductor crystal), where they are finally absorbed and converted into heat. Taking all power losses into consideration, the total fraction of photons with respect to a known power level input that is emitted by the LED is known as the extraction efficiency η_1. Currently, the extraction efficiency of HB-LED is around 50%, which is relatively much lower than other functional stages of energy conversion, and therefore, it is the most influencing factor affecting the overall efficiency of the LED lamps [7].

In Stage 3, the blue light carrying power of $P_{\text{opt-b}}$ is converted into white light by the phosphor with a conversion power efficiency of η_2. The conversion power loss is related to the quantum efficiency and absorption characteristics of the phosphor materials and is influenced by the trapping and absorption of the photons' energy, which is eventually converted into heat by the phosphor material of the LED. Currently, many commercially available phosphor materials are of good performance, with a conversion efficiency η_2 that is usually higher than 90% [8]. The optical power of the emitted white light, $P_{\text{opt-w}}$, is given by Eq. (6.3):

$$P_{\text{opt-w}} = P_{\text{opt-b}}{}^{*}\,\eta_2 \tag{6.3}$$

Finally, the power of white light emitted from the LED passes through Stage 4, which is the lamp cover where the white light is scattered into the ambient. For this stage, the lamp covers, which, in the process of scattering the light, partially trap photons within the covers, converting them into extra heat, thereby incurring an additional form of optical power loss. Thus, the final optical output power of the light emitted to the ambient in terms of the efficiency η_3 can be expressed as $P_{\text{opt-a}}$ by Eq. (6.4):

$$P_{\text{opt-a}} = P_{\text{opt-w}}{}^{*}\,\eta_3 \tag{6.4}$$

There are many types of LED lamps in the world today. In principle, they all have power loss similar to the examples described above. The key to increasing the efficiency is to control the ambient temperature that is within the junction temperature of the LED by having good thermal management in the LED lamp. An efficient thermal management prolongs the life of the LED and the lamp driver.

Having explained the conservation of the optical power from the LED light source to the LED lamp, it is also good to know how the lamp is fixed in the luminaire. There are many types of interior lighting luminaires. A few home applications are discussed here. Figure 6.6a illustrates a luminaire for kitchen applications where the retrofit LED lamp is used for illuminating the workspace. The power of the lamp is shown in Figure 6.6b at 9 W with 902 lm, which is equivalent to a 75 W conventional incandescent lamp. This means that the LED lamp has energy savings of 88% in

(a) (b) (c)

Figure 6.6 (a) Retrofit lamp in the kitchen luminaire. Source: Courtesy of ams OSRAM GmbH. (b, c) LED lamp from Westlite showing energy savings, power, and lifetime.

comparison to conventional incandescent lamp. Also illustrated in Figure 6.6b is the lifetime of the LED lamp at 15,000 hours.

6.1.2 Hospitality Lighting – Architecture Lighting

The LEDs are extremely attractive in the field of architectural and artistic lighting because of their advantages as light sources. To name a few, they are compact, which come in luminaires and lines with different shapes and sizes, electrical power, and lighting characteristics. Besides that, the light generation principle of the LEDs itself leads to the opportunity to obtain different colors of emitted light and creates an attractive night vision of the architectural objects.

Other advantages are the LED's power (luminescence) and durability. Some LED lamps reach an average of fifty thousand hours, and in some cases, they reach close to hundred thousand hours with high energy efficiency [9]. These advantages are useful for luminaries that require very little maintenance. An example is the high ceiling of a ballroom or conference room.

An example of the interior design of the conference hall of St. Giles Wembley's Hotel in Penang, Malaysia, as shown in Figure 6.7, demonstrates creativity and elegance in lighting design to illuminate the large space while at the same time retaining the aesthetic value. This was done by having many downlights spread throughout the hall ceiling at the center of the hall and many small LED lights in the chandelier illuminating the whole hall. Inside the circular decorative frame, many linear LED lights are installed to illuminate the whole hall. The brightness of the LED lights can be controlled individually. During the presentation, the hall light was dimmed, and the speaker's corner was lighted slightly. As the ambient light is a dimmer, the projector image is clearer for the audience. Hence, as a whole, the LED lighting system is much easier to use for dimming and lighting the usable space in the hall.

Figure 6.7 Example of interior lighting in the conference hall of St. Giles Wembley's Hotel, Penang, Malaysia (December 2020).

6.2 LEDfit Lighting

The LEDfit lighting covers a wide variety of lighting similar to retrofit lighting. In this book, we cover only interior lighting, outdoor lighting and horticulture. They are residential lighting, street lighting, architectural lighting, and horticulture. These three lightingss markets have a significant impact on the LED industry in terms of business and artistic values that an LED can provide compared to the conventional light source.

6.2.1 Residential Lighting – Living Room Down Lighting

An example of living room lighting is home lighting. For a small scale of the open space, square-shaped LED downlight can be used across the living room as a lighting option. Figure 6.8 shows an example of square-shaped LED downlight. It gave good lighting coverage throughout the living room. It is not fanciful but provides enough light for decent illumination in the living room. This LED lighting is compact and flat on the surface of the ceiling. It is different from the conventional retrofit LED lamp type of downlighting, where it avoids special reflector recesses on the ceiling. Such recesses on the ceiling look like outdated architectural designs, where consumed space and lacks aesthetic value. This square LED downlight lighting, on the other hand, has its reflector inside the compact lighting design. Hence, no further recesses are needed to accommodate the light on the ceiling. The lifetime of LED is now up to 50,000 hours, so we don't even need to worry about changing the LED light for a very long time.

Figure 6.8 Retrofit square-shaped LED downlight in the living room.

6.2.2 LED Street Lighting

The LED street lighting system has to mimic some of the conventional systems based on high-pressure sodium (HPS) lamps or high-pressure metal halide (MH) systems. However, the old street lighting technologies have some issues. For example, the maintenance of an outdoor lighting system is expensive and can affect road users. Hence, LED comes as a perfect fit for this application, where LED with a longlife span reduces street light maintenance to a very minimum. The conventional lighting system life span is roughly 15,000 hours for HPS and 20,000 hours for MH [10]. This is nowhere near LED-based light at roughly 50,000 hours [10]. Some reports stated that the LED street light lifespan is up to 100,000 hours [11]. These advantages, combined with energy savings of 30–40% make LED-based outdoor lighting an irresistible choice [12]. Other issues with conventional lighting are that the light distribution is difficult to control as it emits in all directions, as shown in Figure 6.9. That has consequences such as glare, nonuniform light patterns, light pollution, and energy waste . Old lighting technologies limit both eye comfort and visual discrimination abilities of car drivers and pedestrians. In this context, LED street lighting introduces a new concept of adaptive street lighting where light is more focused on the street and energy is saved because it is not wasted on the sky or the side of the road.

The European standard (ENI3201:2015) for road lighting quality characteristics gives values for average road surface luminance and overall and longitudinal luminance uniformity [11]. These values are based on photopic photometry. However, during the nighttime, the visual conditions are usually mesopic. Therefore, light sources that are more effective under mesopic conditions can be used to reduce the luminance on the road surface while providing the same visibility. Thus, mesopic design has the potential to save energy [11].

When we talk about satisfying the requirements of road lighting system, one must understand mesopic photometry. The human eye is not equally sensitive to all wavelengths of light. The sensitivity of the eye to different wavelengths is described by the relative spectral sensitivity function. Mesopic vision relates to lighting levels

(a) (b)

Figure 6.9 Comparison between conventional street lighting and LED street lighting. (a) Street light in one of the housing estates in Penang, Malaysia, illuminated by yellowish-orange high-pressure sodium luminaries. (b) Street lights in one of the housing estates in Penang, Malaysia, illuminated by cool white LED luminaries at 5700k CCT.

between photopic and scotopic vision. Under mesopic conditions, both rods and cones tissues in the retina are active. The mesopic spectral sensitivity is not constant but varies with light level and viewing conditions due to the distribution of rods and cones on the retina. The CIE system for mesopic photometry is based on a linear combination of photopic and scotopic spectral sensitivity functions, as shown in Figure 6.10c. The upper luminance limit of the mesopic system is $5\,cd/m^2$, and the lower luminance limit is $0.005\,cd/m^2$, as illustrated in Figure 6.10c [13].

The International Commission on Illumination (CIE) has published The Technical Report 191: 2010 Recommended System for Mesopic Photometry based on Visual Performance [CIE2010]. The new CIE mesopic photometry is valid between 0.005 and $5\,cd/m^2$, i.e. it covers luminance encountered in outdoor lighting. The new mesopic system provides for the first time the means to evaluate the lighting in terms of an internationally accepted system of mesopic photometry [13].

In Figure 6.11, an example of an LED-based outdoor luminaire for street lighting is shown. The LED luminaire incorporates three optical elements: Total internal reflective (TIR) reflector, a LED base, and a cover sheet. Each TIR reflector collimates and recycles the light that is emitted from the LEDs. This TIR reflector improves the overall efficacy of the luminaire. And the cover sheet, on the other hand, uniformly distributes light only within the street boundaries.

The investment required for LED base lighting over the entire product life is very competitive compared to conventional lamps. The initial cost of an LED lamp is much higher than that of an HPS or MH lamp. However, the total cost of using LED street lamps over long term is cheaper compared to conventional lamps.

The beam pattern of the distribution curve can also be changed according to the demands. To control the light distribution rationally, a rectangular beam pattern, as shown in Figure 6.12a,b, is perfectly fit for road lighting. The edge of the beam

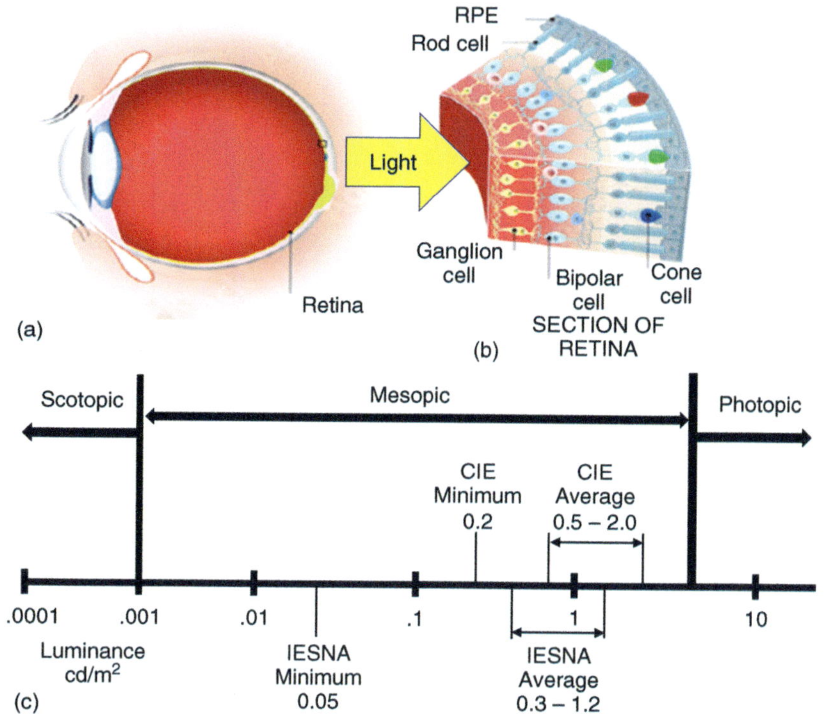

Figure 6.10 (a) Cross-section through a human eye. Source: Adapted from Schubert et.al. [3]. (b) Schematic view of the retina, including rod and cone light receptors. Source: designua/Adobe Stock. (c) Typical ambient light levels compared to photopic, mesopic, and scotopic retinal illumination and visual function. Source: Adapted from Sahana et al. [13].

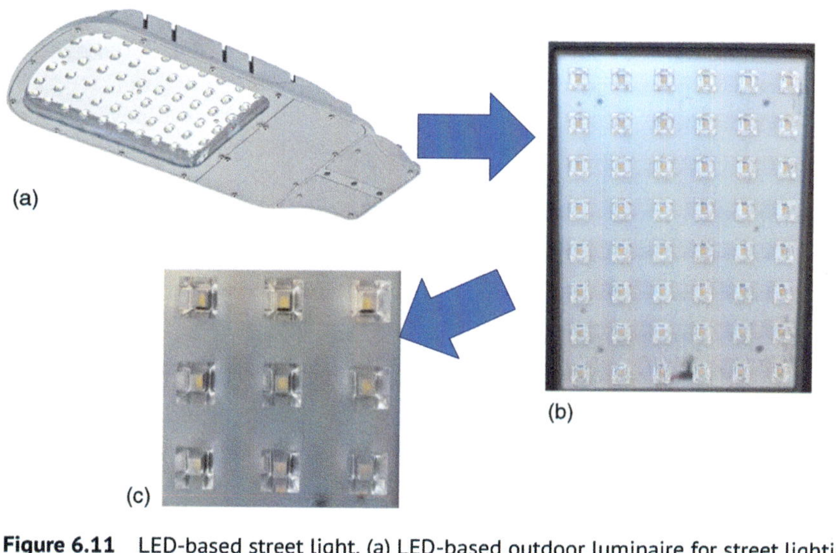

Figure 6.11 LED-based street light. (a) LED-based outdoor luminaire for street lighting. (b) LED street lighting in close view of the LED. (c) Total internal reflective (TIR) reflector.

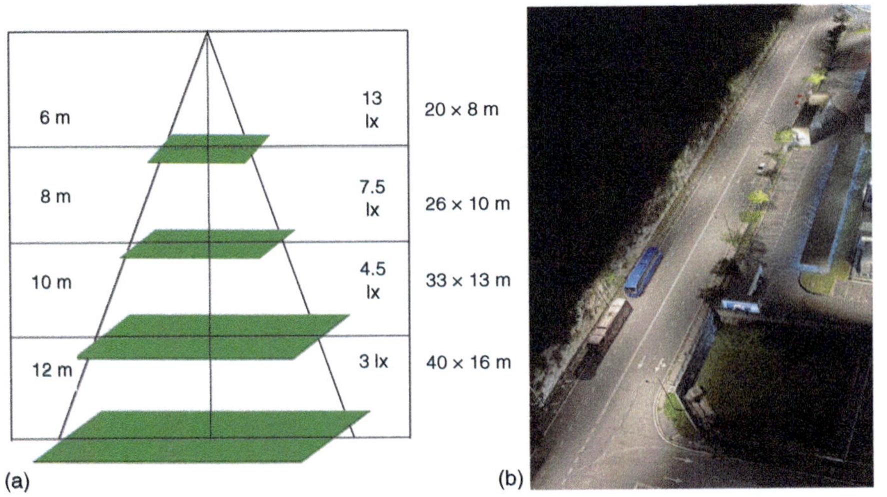

Figure 6.12 (a) Street light illumination distribution against height. (b) Street lighting distribution on the roads as viewed from high-rise buildings.

pattern is very clear, with no glare out of the effective radiation region, which will not cause any light pollution. Hence, satisfy the requirements of road lighting.

The purpose of road lighting is to make people, vehicles, and objects on the road visible without causing discomfort to the driver. The European standard for road lighting gives values for illuminance and luminance and their distribution on the road surface. Furthermore, the standard gives measures of the loss of visibility caused by the glare of the luminaires of a road lighting installation [13]. The standard also defines and describes the conventions and mathematical procedures for calculating the photometric performance of road lighting installations. The range of the average road surface luminance recommended in the standard is between 0.3 and 2 cd/m^2, which is in the mesopic region. However, all the lighting quantities used in the standard are based on the photopic $V(\lambda)$ function [13].

Light sources that have high Scotopic/Photopic (S/P) – ratios and have spectral output in the blue wavelength region are myopically more effective. The higher the S/P-ratio is, the better the light source is in terms of mesopic design. The spectral power distribution of light defines its color characteristics, described with the correlated color temperature (CCT) and the general color rendering index (CRI). There are, however, shortcomings of the CRI when applied to LED light sources due to their peaked spectrum. The CIE technical committee TCI-69 Color Rendering of White Light Sources is investigating this issue while writing this book. The effect of light color has been studied in indoor lighting.

The other issue with street lighting is the glare issue. Glare is related to phenomena where visual perception is hampered or even impossible. Glare has three aspects: dazzle, physiological glare, and psychological glare [14]. Dazzle, or blinding glare, occurs when the intensity of the light stimulus rises above the upper limit of the sensitivity area of the visual system. The stimulus obstructs relevant perception. Dazzle occurs, for example, when driving on a wet surface against a low sun or leaving a tunnel in daylight. Physiological glare, also called the disability glare, occurs when one or more glare sources occur in the field of vision. A light veil is formed in the whole field of vision, which reduces contrast and the visibility of the target. Psychological glare, also called discomfort glare, occurs when glare sources in the field of view (FOV) cause disturbing effects and discomfort in vision without reducing the visual performance.

Having said about the impact of the street light on the visual performance of the road user, now we can summarize that replacing the street light with LED lighting has great benefits. Energy savings are huge. The initial cost of replacing LED lighting luminaires can be expensive; however, looking at total cost, including energy savings and low maintenance costs, certainly makes LED lighting stand out. Besides that, the mesopic effect of LED is much better compared to conventional lighting.

6.2.3 Exterior Architectural Lighting

One of the amazing contributions of LED lighting is architectural lighting. The colors and the art of external illumination have substantially influenced the creation and the perception of architecture and the nighttime urban environment. Every object, part of the perceived environment, is associated with its inherent attributes, some of which are critical for its recognition and others influence the estimation of its properties. The lighting specialists are responsible not only for choosing a light that makes objects visible but also for enhancing them attractively. In order to achieve good architectural lighting, attention should be paid not only to the light sources used but also to the materials used for the different surfaces and their interaction with light. The reflectance and color of the architectural elements are of great importance. Using an optimal combination of the light spectrum of the source of light and reflective and color characteristics of the surfaces of the architectural elements (in terms of the interaction of light and matter) will lead to the best lighting results. Based on the effects obtained, the best night vision of the objects becomes reality.

How these can be achieved? The architectural style is determined by the tectonics of space, the materials used, the constructive methods, the method of shaping, and the semantics of the images. All these are to some extent due to the new approaches to the design of the light-color medium. Even if one defines the architectural style only by a set of external attributes, then, in this case, artificial light can create many new images.

The game of light in religious buildings has always been a powerful means of influencing people's feelings. Figure 6.13 shows the Kek Lok Si temple in Penang, Malaysia. The artistic and architectural aspects of lighting design gave images that influenced the feeling of the parishioner or even visitors.

The novelty of the approach to the style-forming role of the lighting design is to create a diverse and dynamic environment, using the distribution of artificial light in form of LED light source.

It was there that historically people formed receptions, which then began to be used in holy places and theaters. Now, these forms of architectural lighting have gone

Figure 6.13 Day and night views of Kek Lok Si Temple, Penang, Malaysia.

Figure 6.14 Outdoor lighting – Stadium San Mames, Bilbao, Spain. Source: Courtesy of ams OSRAM GmbH.

to much wider places, like modern buildings and sports complexes like stadiums. The evening illumination of buildings was invented with bright images to impress and stand out. Modern skyscrapers in Shanghai, Guangdong, New York, Tokyo, and many more cities are lighted with colorful lights. Not a single iconic object is being built today without the development of evening architectural lighting. Same goes for stadiums. The illumination indoors and outdoors is spectacular, with beautiful architectural lighting. Figure 6.14 shows the image of famous architectural lighting of San Mames Stadium in, Bilbao, Spain. These buildings are not only beautiful in the evening; they are actually, an attraction for tourism. The brilliant, colorful, and dynamic lighting is amazingly beautiful.

6.2.4 Horticulture Lighting Application

The use of artificial light for plant growth and development purposes has been known for long time. Particularly in the last few decades, lighting technologies, such as fluorescent (FL), HPS, MH, and incandescent (INC) lamps, have started to be implemented for plant cultivation and research. However, not much progress has been made, due to some limiting factors. Light is the most important source of energy for photosynthesis; therefore, the lighting environment surrounding the plant canopy can differently influence plant growth and development. The conventional artificial lighting sources (e.g. FL, HPS, MH, and INC lamps) can present several drawbacks, as they are neither spectrally optimal for crops nor energetically efficient, furthermore releasing a large amount of radiant heat. These drawbacks are known, and an alternative artificial light that overcomes these drawbacks is the LEDs, which are perfectly suited to horticulture application needs and are increasingly replacing the conventional lighting technologies in many indoor and protected environments, resulting in the rapid technological evolution in the horticultural lighting industry.

Horticultural LED luminaires represent solutions that are more environmentally friendly and economically favorable than conventional lighting while having safer management and disposal practices. In contrast, with traditional HPS lamps, which convert only 30% of the energy into usable light, with significant radiation losses in the form of heat, LED lighting sources often convert about 50% of the electricity into light, resulting in an economically and energetically better solution [15]. Additionally, due to a nonuniform distribution of solar radiation across world regions, supplemental illumination has been used at higher latitudes countries in greenhouse crop production, allowing improvements in terms of productivity and quality and enabling year-round cultivation. The suitability of LED lighting systems for plant-growing applications is based on their potential features (e.g. small size, durability, long lifetime, and cool emitting temperature) in combination with the advantages offered by the modularity in wavelength selection and light output and the elevated energy conversion efficiency.

LED application for plant growth was first studied in the 1990s, when NASA-affiliated researchers performed much of early work as preparation for the development of plant-based regenerative life-support systems for future Moon and Mars bases [16]. However, at that time, only red (660 nm) LEDs were available. Today, LED lighting systems have experienced wide evolution in terms of physical shapes and designs, waveband color availability, power use reduction per unit of light output, and cost decrease per unit of light output. There are wide variety of LED lights in different wavelengths, powers and prices for horticulture industry to choose from for their plant cultivation. Each plant type has its own lighting requirement. Hence, scientists and horticulturists have done much research in this area to select the right light wavelength for specific plants to have optimum photosynthesis and photomorphogenesis.

Light is a source of energy for photosynthesis as well as a source of signals or information activating photomorphogenesis and other physiological processes such as secondary metabolite production in plants. The wavelengths of photosynthetically active light (400–700 nm) and physiologically active light (300–800 nm) overlap, so that photosynthesis and photomorphogenesis are often concurrent [17]. Both are photochemical reactions, and the amount of light received by the plants is measured in units of moles (1 mol = 6.03×10^{23} photons), not in joules (energy). Photosynthesis requires a higher energy requirement, whereas photomorphogenesis has a lower energy requirement [17].

6.2.4.1 Photosynthesis

Photosynthesis is a chemical reaction that takes place inside a plant where the plant takes in carbon dioxide (CO_2) from air, water from ground, and light from the sun or artificial light and converts them into glucose ($C_6H_{12}O_6$) and oxygen (O_2). The

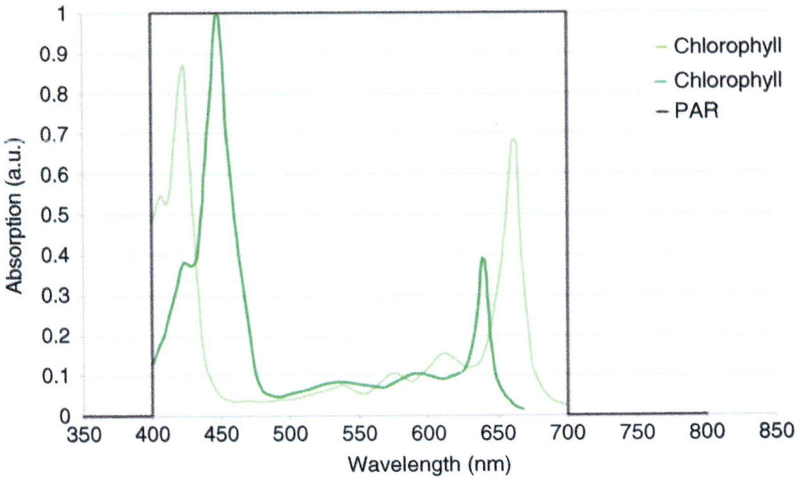

Figure 6.15 Photosynthesis is fueled by blue or red light – peaking at 460 and 680 nm, respectively. Source: Courtesy of ams OSRAM Group [19].

chemical reaction can be described by Eq. (6.5) [18]:

$$6CO_2 + 6H_2O \xrightarrow{\text{light}} C_6H_{12}O_6 + 6O_2 \tag{6.5}$$

The visible light is captured by the carotene and chlorophyll pigments in leaves. Photosynthetic rates are highest in 2 bands: red light, with some activity in the blue–green wavelengths as shown in Figure 6.15. These wavelengths are collectively known as photosynthetically active radiation (PAR).

6.2.5 Photomorphogenesis

Photomorphogenesis is an organism's response to information present in light environment. It is the process by which plants grow and develop in response to light signals. This process is mediated by a sophisticated network of photoreceptors, as shown in Figure 6.16, among which phytochromes play a key role in photomorphogenesis.

In Figure 6.17, recommended LED emission spectrum for plant growth is shown. For plants to grow, it is recommended to use a photon flux at ratio of 10% deep blue (450 nm) and 90% hyper red (660 nm) based on OSRAM Opto Semiconductors findings. However, there is much new research in horticulture that shows some variation in the flux ratio from plant to plant. It all boils back to the plant and desired outcome of the crop output.

The pigment phytochrome in the plant is sensitive to red to far-red light and acts as an environmental sensor to measure daylength. The phytochrome system controls several aspects of seedling phenology, such as seed germination and bud set [16].

The other photoreceptors are cryptochromes, which are sensitive to blue light and long-wavelength UV (UV-A)[16]. Cryptochrome receptors control stem elongation, leaf expansion, circadian rhythms, and flowering time. Blue and UV-A light are

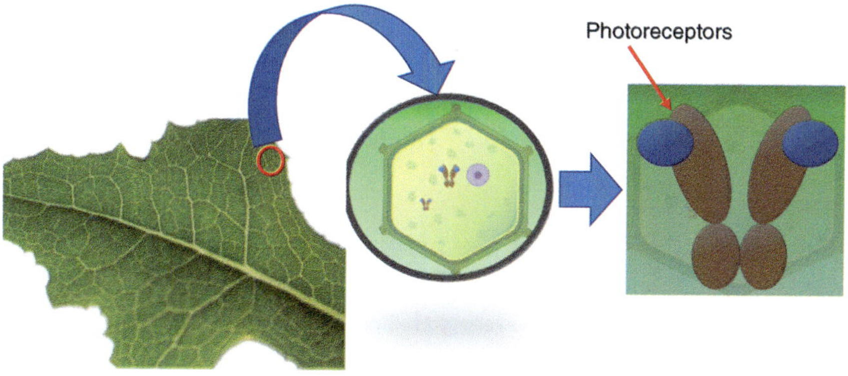

Figure 6.16 Plant photoreceptors.

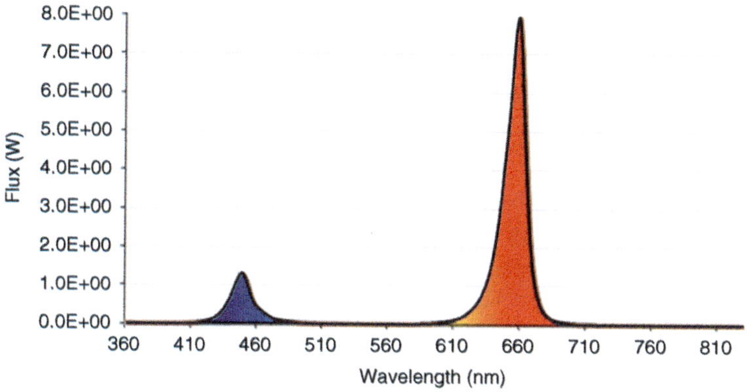

Figure 6.17 LED emission spectrum recommended for plant growth by supplemental lighting. Source: Courtesy of ams OSRAM GmbH [19].

important to normal morphological development, particularly regarding branching and shoot sturdiness. Conceptually, the phytochrome system can be viewed as a light switch. Under predominantly red light, the switch is "on," and cell growth occurs as fast as the light intensity permits. However, when far-red light predominates, the switch is turned "off" and growth stops as plants transition into dormancy.

6.2.5.1 Impact of LED Light on Horticulture Industry

The science of plant growth has been briefly discussed in previous subchapters. In this subchapter, we further enlighten the impact of LED light on horticulture industry.

It's already a common knowledge nowadays that LED is used as an artificial light source in the horticulture industry. The light produced by LEDs can mimic the sunlight, but it can be tuned per the plant's requirements. It affects the plant's growth significantly. The impact of LED on horticulture industry can be summarized in following aspects, and perhaps in more:

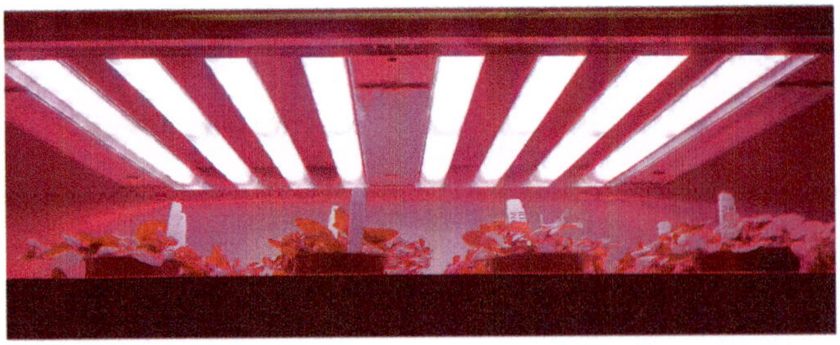

Figure 6.18 Light-emitting diodes (LEDs) used in horticulture are arranged in arrays designed to produce light of specific wavelengths. Since they do not radiate heat, can be located within plant canopies. Source: Courtesy of ams OSRAM GmbH.

LED lifespan: The usable life of LED units is significantly longer than that of traditional artificial light sources used in horticulture, ranging from two to three times better than fluorescent or HID lamps to a 50-fold increase over typical incandescent lamps. Unlike traditional lamps, LEDs do not "burn out"; instead, they gradually dim and should be replaced once they dim to 70%. Since the LED light's lifetime is long, less maintenance work on the LED light source is required. Hence, this significantly reduces costs and increases profitability for horticulture industry.

Energy efficiency: As measured by radiated power output (lumens) divided by the electrical power input (watts), LED units are very efficient, especially when compared to traditional incandescent bulbs. The energy efficiency of LED lights continues to improve every year. The latest report stated that the LED efficiency had already hit more than 200 lm/W [20]. This means less energy use for plants, thereby saving money. Studies also show that LED lights produce more plant output compared to conventional light sources [16, 19].

Radiant heat: LEDs produce almost no radiant heat and so can be positioned close to plants, ensuring maximum light interception. Hence, reduce photon loss and save money.

Custom lighting: LEDs produce light in a very narrow wavelength range, so units can be designed to produce light of desired wavelengths, or combined to generate white light. LED arrays of blue and red light that increase photosynthesis can be positioned within crop canopies and close to plants, as shown in Figure 6.18. This further enhances the absorption at lower leaves.

Plant productivity: Many studies show that LED lighting produces relatively more crop output compared to overhead HPS lighting. There are many benefits of using LED for horticulture; however, there are also some disadvantages, for example, the initial capital investment of LED lighting in horticulture is relatively high. In the long run, LED lighting certainly shows more advantages than disadvantages.

6.3 Summary

LED in general lighting has greatly benefited humankind. LED light has brought down the cost of artificial light generation greatly. LED lights have brought comfort at home and outside the home, like in street lights, buildings, and religious houses. Besides that, LEDs have also helped indoor cultivation produce more vegetables in more affordable and ecologically safe environment.

It is reported that LED lighting will reduce fuel consumption and carbon dioxide emissions from power stations by at least 10% over next 5–10 years.

References

1 Nakamura, S. (1991). GaN growth using GaN buffer layer. *Japanese Journal of Applied Physics* 30 (10A): L1705.

2 Nakamura, S. and Chichibu, S.F. (2000). *Introduction to Nitride Semiconductor Blue Lasers and Light Emitting Diodes*. New York: Taylor & Francis.

3 Schubert, F.E. (2006). *Light Emitting Diode*, vol. 2. New York, USA: Cambridge University Press.

4 Liu, S. and Luo, X. (2011). *LED Packaging for Lighting Applications: Design, Manufacturing, and Testing*. Wiley.

5 Jakovenko, J., Formánek, J., Janíček, V. et al. (2012). High power solid state retrofit lamp thermal characterization and modeling. *Radioengineering* 21 (1).

6 Jordà, X., Perpiñà, X., Vellvehi, M. et al. (2013). Influence of different characterization parameters on the accuracy of LED board thermal models for retrofit bulbs. In: *19th International Workshop on Thermal Investigations of ICs and Systems (THERMINIC)*, 194–199. IEEE.

7 Shen, Y., Shen, Y., Yin, J. et al. (2020). Lead-free, stable, high-efficiency (52%) blue luminescent FA_3 Bi_2 Br_9 perovskite quantum dots. *Nanoscale horizons* 5 (3): 580–585.

8 Zhong, Y., Xia, M., Chen, Z. et al. (2020). Pyrophosphate phosphor solid solution with high quantum efficiency and thermal stability for efficient LED lighting. *Iscience* 23 (3): 100892.

9 Su, D., Casamayor, J.L., and Xu, X. (2021). An integrated approach for eco-design and its application in LED lighting product development. *Sustainability* 13 (2): 488.

10 Jia, Z. (2020). Comparison on lamp characteristics of highway tunnel lighting system. In: *IOP Conference Series: Earth and Environmental Science*, 1–8. IOP Publishing, 052095.

11 Fryc, I., Czyżewski, D., Fan, J., and Gălăţanu, C.D. (2021). The drive towards optimization of road lighting energy consumption based on mesopic vision—a suburban street case study. *Energies* 14 (4): 1175.

12 Kusuma, P., Pattison, P.M., and Bugbee, B. (2020). From physics to fixtures to food: current and potential LED efficacy. *Horticulture Research* 7 (1): 1–9.

13 Sahana, S., Datta, D., and Roy, B. (2020). Estimation of mesopic adaptation luminance under different surrounding lighting ambience. In: *2020 IEEE Applied Signal Processing Conference (ASPCON)*, 173–177. IEEE.

14 Schreuder, D. (2008). Road lighting design. *Outdoor Lighting: Physics, Vision and Perception* 401–436.

15 Morrow, R.C. (2008). *LED lighting in horticulture. HortScience* 43 (7): 1947–1950.

16 Mitchell, C.A. and Sheibani, F. (2020). LED advancements for plant-factory artificial lighting. In: *Plant Factory* (ed. T. Kozai, G. Niu, and M. Takagaki), 167–184. Elsevier.

17 OSRAM (2021). *LEDs for horticulture lighting – OSRAM*. Online Horticulture Lighting Application Notes.

18 Britannica (2021). *Photosynthesis*. Online Document – Photosynthesis, Formula, Process, Diagram.

19 Stefanie, R. and Alexander, W. (2021). *LED for Horticultural Lighting Application*. OSRAM Opto Semiconductors.

20 Taki, T. and Strassburg, M. (2020). Visible LEDs: more than efficient light. *ECS Journal of Solid State Science and Technology* 9 (1): 015017.

7

Quantum LEDs

7.1 Quantum LED as the Alternative to Organic LED

As we already learned in-depth about the fundamental working principles of LEDs, their manufacturing, and their usage in many fields, we understood that the LEDs provide superior luminous efficiency, low cost-high volume manufacturability, and extended life cycle. The low power consumption of LED panels makes them lucrative for domestic usage; however, the color filters used absorb large part of the emitted photons, which eventually reduces the light use efficiency of LED display units. Theoretical value of the LED display unit's light use efficiency is still <2.8%, which indicates more power is needed to sustain expected light brightness [1]. Here, two main technologies come into play: the organic-LED and the semiconductor-LED. Both come with their own advantages and disadvantages, where OLED can be manufactured on large scale and fitted into rigid or flexible substrates; however, the short lifetime and inferior light brightness make them less favorable. The semiconductor-LED, on the other hand, is superior compared to OLED in many ways but comes with price where you need three different RGB LEDs to provide color mixing. To produce tricolor in single wafer was deemed tedious from manufacturing point of view besides their cost competitiveness. To address these gaps, a new territory in LED has arisen, which is quantum dot LED (QD-LED). Quantum lighting in principle comes from the blue light excitation on the quantum dots, which causes pure basic light emission with the least color interference and low light intensity loss and eventually increases the color sharpness and quality [2–4].

7.2 Fundamentals of Quantum Dot

The term quantum dot arises from the physical phenomenon known as quantum confinement, which is the redistribution of electronic energy when the size of the material is smaller than a fundamental length scale of semiconductors, known as exciton Bohr radius. Excitons are quasi-particles formed as a result of light absorption by a semiconductor crystal or molecule, i.e. the electric dipole with a hole in the valence band (or HOMO of a molecule or cluster) and electron in the conduction band (or LUMO of a molecule or cluster) created as a result of light absorption.

LED Packaging Technologies: Design, Manufacture, and Applications, First Edition.
Luruthudass Annaniah, Mohamed Salleh M. Saheed, and Rajan Jose.

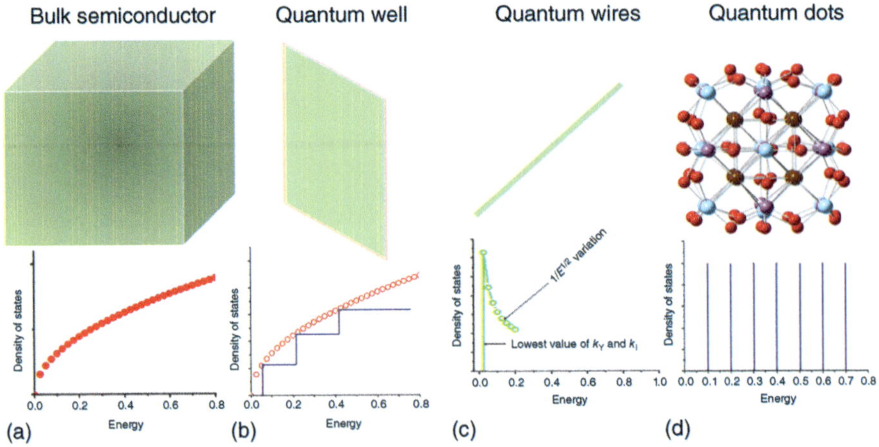

Figure 7.1 Summary of the changes in the density of states of electrons in the (a) bulk crystal and when the size of the semiconductor becomes lower than the exciton Bohr radius, i.e. (b) quantum wells, (c) quantum wires, and (d) quantum dots. Insets of each figure show a structural diagram of each material's architecture.

There are no exciton constraints on the surface of the bulk semiconductor where it is formed, i.e. it is free to move throughout the semiconductor. However, if the size of the exciton is larger than the material it holds, the electron energies are to be redistributed to allow its existence in the material. This phenomenon is known as quantum confinement. Consequently, there are one-, two-, and three-dimensional confinements, as shown in Figure 7.1. One-dimensional confinement occurs when one of the dimensions of the semiconductor is smaller than its exciton Bohr radius, such material is called quantum well structures (Figure 7.1) and is extensively used in semiconducting lasers. Similarly, a quantum wire exhibits two-dimensional confinement. If the physical size of the semiconductor is smaller than the exciton Bohr radius, such materials are called quantum dots. Given the size of the exciton Bohr radius, which is a few to few tens of nanometers, all quantum dots fall into the category of nanostructured materials. Equations (7.1)–(7.4) detail the electron density of states of bulk crystals to quantum dots [5].

$$\text{Bulk crystal} \quad \rho' = \frac{1}{2\pi^2} \times \left(\frac{2m}{\hbar^2}\right)^{\frac{3}{2}} \times \sqrt{E} \tag{7.1}$$

$$\text{Quantum wells} \quad \rho' = \frac{mE}{\pi\hbar^2} \tag{7.2}$$

$$\text{Quantum wires} \quad \rho' = \frac{2}{\pi} \times \sqrt{\frac{2m}{\hbar^2}} \times \frac{1}{2\sqrt{E}} \tag{7.3}$$

$$\text{Quantum dots} \quad \rho' = \delta(E - E_{(m,n,o)}) \tag{7.4}$$

In the above equations, ρ' is the electron density of state, m is the conduction band effective mass, E is the unconfined energy, $E_{(m,n,o)}$ is the confined energy with their respective quantum numbers, and $\hbar = h/2\pi$, where h is the Planck's constant. Variation in electron density of these structures is also drawn in Figure 7.1.

Equation (7.1) would imply that the electron density of state of a quantum dot is a delta function, $\rho' = 1$ when $E = E_{(m,n,o)}$ and $\rho' = 0$ when $E \neq E_{(m,n,o)}$. In other words, the electron density of states is discrete and only exists at certain conditions of the quantum conditions, like that of an atom. Due to such electron density of states, quantum dots are often called "artificial atoms." One would observe from Figure 7.1 that quantum dots consist of only a few tens to hundreds of atoms.

The redistribution of electron energy is manifested in the material as a modification of the electronic bandgap expressed in the form [6]:

$$E_{1\mathrm{sh},1\mathrm{se}}(R_{\mathrm{av}}) = E_{\mathrm{g}} + \frac{1}{2}\left(\frac{h\pi}{2R_{\mathrm{av}}}\right)^2 \times \left(\frac{1}{m_{\mathrm{e}}} + \frac{1}{m_{\mathrm{h}}}\right) - \left(\frac{1.8e^2}{\varepsilon R_{\mathrm{av}}}\right) \tag{7.5}$$

where $E_{1\mathrm{sh},1\mathrm{se}}$ is the energy of the first excitonic peak, which can be determined from the absorption spectrum, R_{av} is the average particle size, E_{g} is the band gap energy, h is the Planck's constant, m_{e} is conduction band effective mass, m_{h} is valence band effective mass, ε is the dielectric constant, and e is the electronic charge. The size-dependent tuneability of absorption and emission of a compound semiconductor is shown in Figure 7.2, and this property enabled quantum dots for applications in life science, lasers, LEDs, etc.

In the early 1990's, the QD was researched extensively and many methodologies such as chemical synthesis were developed to produce wavelength emissions to be narrow, highly efficient, and controllable by tuning the chemical compositions and QD sizes. Briefly, the quantum dots are synthesized via nucleation and growth method via injection of molecular precursors kept at room temperature into solvent maintained at temperatures in the 300–400 °C range in a controlled atmosphere [8]. Such chemical reaction produces a conformity and homogeneous QD sizes as desired. This is achieved by controlling solution temperature, reaction time, precursor ratio, and concentrations [2, 9]. These advancements in QDs precise preparations enabled the researchers the freedom to control light's wavelength emission range and narrow distribution.

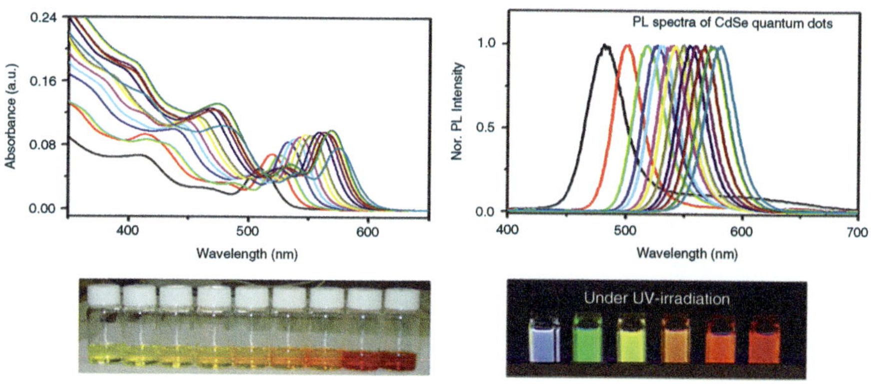

Figure 7.2 Absorption and emission spectra of CdSe quantum dots synthesized via hot emission technique. Source: Adapted from Jose et al. [7]/American Chemical Society.

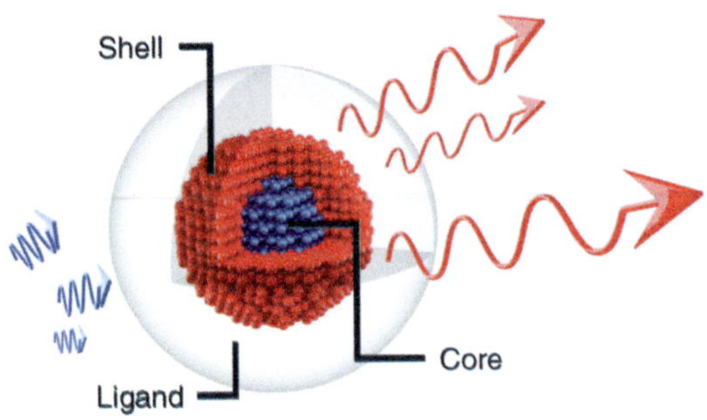

Figure 7.3 Quantum dot with core inside the shell with ligand overcoated. Source: Courtesy from ams OSRAM GmbH.

Cadmium-selenium (CdSe) is the simple and most studied QD, crystallized in cubic zinc blende and hexagonal wurtzite structures. As shown in Figure 7.1, most atoms in QDs are on the surface, and consequently, there are dangling (or unsaturated bonds). These dangling bonds make the surface energy and reactivity of QDs extremely high, and consequently, they become unstable. This instability is efficiently managed in many ways, such as via surface coating with a wider bandgap material than the parent QDs or passivating the surface atoms with molecular ligands, as shown in Figure 7.3. Therefore, most QDs for optoelectronic applications consist of three main parts: core QDs, outer shell of a wider bandgap material, and the surface ligands. With radiative recombination between hole–electron as shown in Figure 7.4, light is generated within the nanocrystal core, and the nanocrystal size and composition determine the emission wavelength [3]. With these facts, the important parameter for producing high-quality light output was the precise control over homogeneity of nanocrystals.

Over the years, extensive research was carried out to enable maximum light output, core-shell distribution as well as material stability against environment. The material optimization comes from many routes, such as reducing the core and shell lattice mismatch, core-to-shell thickness ratio, and enhancing surface ligands. These were deemed important to enable the process integration of QDs into existing tools and scale up the manufacturability of the QDs. However, with the enhancement of QDs quantum yield, there are several issues to be addressed, such

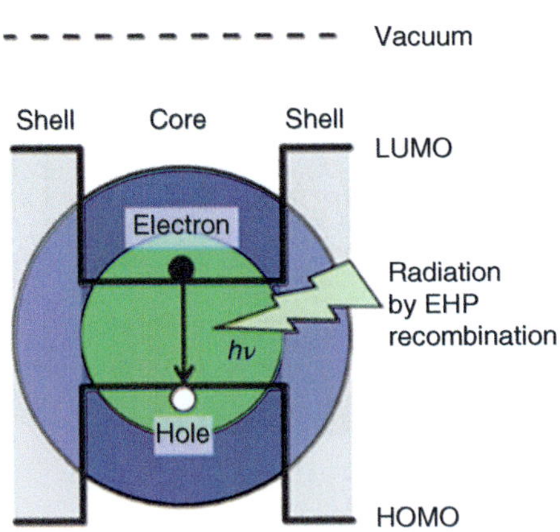

Figure 7.4 Radiative recombination confined within the nanocrystal core. Source: Adapted from Bozyigit and Wood [3].

as the use of heavy metals such as cadmium, tellurium, and selenium, which were classified as non-environmental and hazardous to human health. This disadvantage poses the biggest stone wall for many LED manufacturing giants from deploying QDs in their products.

7.3 Quantum Dots in LED

Maximum LED light output at specific wavelengths requires maximum quantum yield from the QDs. In order to achieve this, non-radiative emission needs to be reduced where the recombination rate occurs via trap sites. In recent year, an important work to produce more environmentally friendly core shell and better stability against ambient conditions were produced.

Figure 7.5 shows the InP core with ZnSe shell synthesized in-situ in HF solvent, which was mainly to etch away the oxide species from InP core. The non-radiative emission was further reduced by increasing the shell thickness, which eventually results in QD stability over time of 100 cd/m^2 for 1,000,000 hours and improved EQE of 21.4%, record high for non-Cd QDs [10].

The challenge to produce non-Cd QDs continued for the green and blue emission QDs. The quantum yield for the green QDs was achieved by employing Zn-oxo compound to produce uniform InP core sizes and reduction in InP core defect density during the synthesis, as pictured in Figure 7.6. Such method is able to produce QY of 95% and narrower green wavelength compared to 84% for the red wavelength, respectively [11]. This compositional alloying is further explored by also inserting

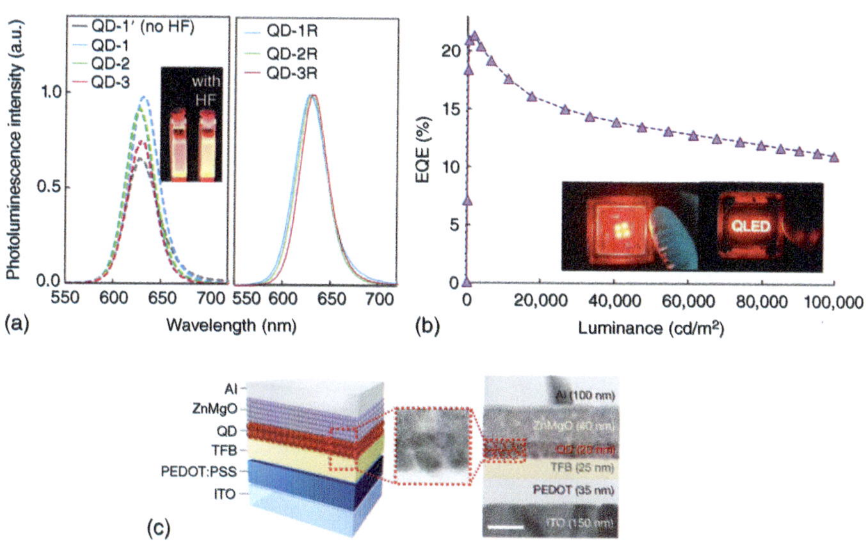

Figure 7.5 (a) Red-colored QDs with and without HF modification. (b) EQE vs luminance of red QD LED. (c) QD LED device structure. Source: Bang et al. [10]/John Wiley & Sons/CC BY 4.0.

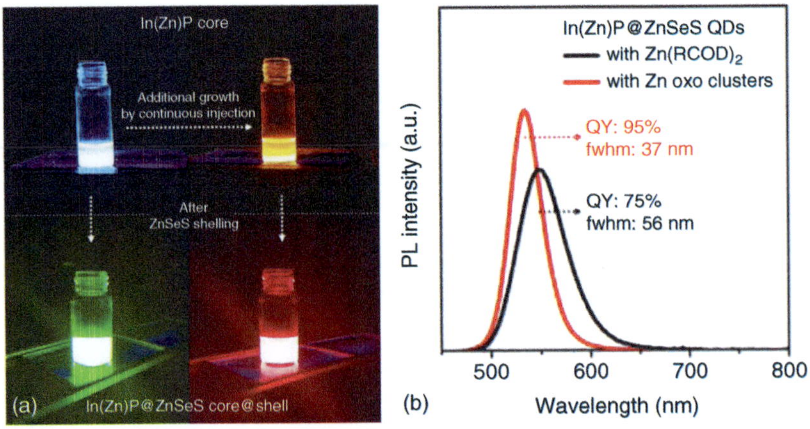

Figure 7.6 (a) InP core after being coated with ZnSe shell, which produces green and red QDs. (b) Maximum QY 95% from Zn oxo compound treated compared with non-treated. Source: Kim et al. [11]/with permission of American Chemical Society.

intermediate layer between core and shell. The function of the intermediate layer is to reduce the lattice mismatch as shown for ZnS/ZnTeSe/ZnS, which gives out an increase in QY to 85% and narrow blue wavelength of 23 nm [12]. With various types of techniques and material engineering being carried out, there are still many gaps to be overcome, such as the use of rare and costly minerals like indium, gallium, and others [13–15]. We will discuss methods of reuse and recycling of these materials after the end of life in Chapter 9.

7.4 Quantum LED Structures

With all the superior properties we discussed above, the QD-based LED needs to be realized by carefully integrating it within the semiconductor material. The usual layers used in the LED will be used here, where anode and cathode electrodes are at both potential sides, and QDs will be the emissive layer (EL), sandwiched in between hole injection layer (HIL) and electron transport layer (ETL). In some cases, the hole transporting layer (HTL) is also integrated together with HIL. In the market, there are two main types: the typical vertical structure and the inverted structures. In general, Figure 7.7 shows the simple QD LED structure, during the forward bias, the electrons are carried via ETL and the holes are carried via HTL/HIL, which are later injected into the EL QDs, photon emissions were produced via Auger radiative recombination. Depending on the QDs core-shell structure and material, various wavelengths of light emissions can be produced, as discussed in Section 7.3. Figure 7.7b shows the commonly used indium-titanium-oxide (ITO) and aluminum (Al) as cathode and anode materials. Many studies also indicate zinc oxide as superior ETL compared to TiO_2. This structure is deemed as favorable one, as the light emission can easily exit the ITO/glass substrates [16].

Besides the QDs, much research has also been carried out on ETL and HTL materials to further enhance the luminance and stability. In recent studies, ETLs such as

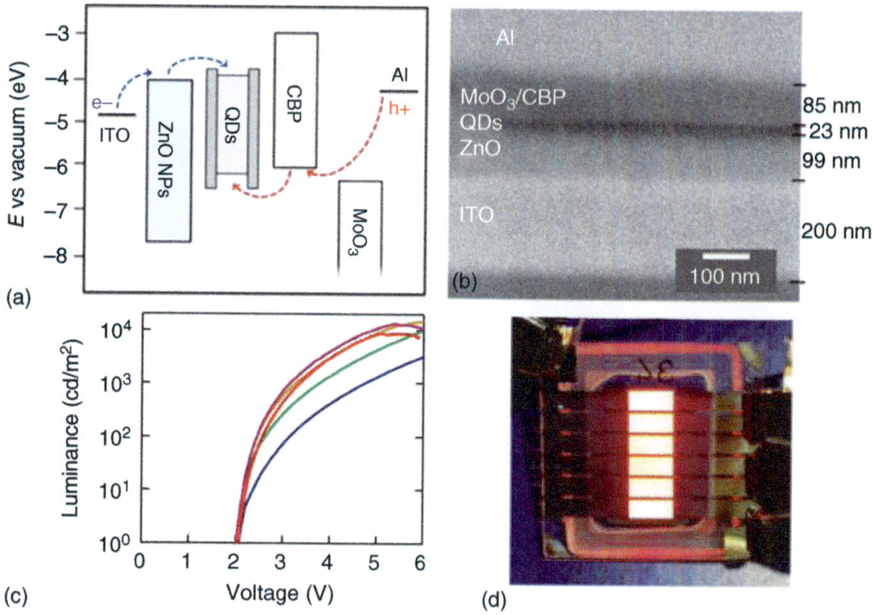

Figure 7.7 (a) Energy level diagram of material used for the QLED. (b) Thickness of QLED from the SEM cross-section. (c) QLED luminance corresponding to the voltage increase. (d) A photo of QLED device with forward bias of 3 V. Source: Kirkwood et al. [16]/with permission of John Wiley & Sons.

ZnO were doped with different materials such as Al, Ga, Mg, and Li to increase the electron conductivity on the ZnO surface. The Al-doped ZnO produces high brightness at the lowest forward bias voltage [17]. These doping plasmonic materials, such as Ag, Cu, Ti, and Ag in ZnO, were proven to reduce the surface resistance and increase the electron–hole mobility within the interface. The improvements in both ETL and HTL also contribute to higher luminance at lower forward bias voltage. Figure 7.8 shows the Cd-free structure with different selections and doped ETLs and HTLs. The deployment of double HTLs improves the hole injection by reducing the HTL to QDs mismatch plus the HTL doping further enhances the hole injection.

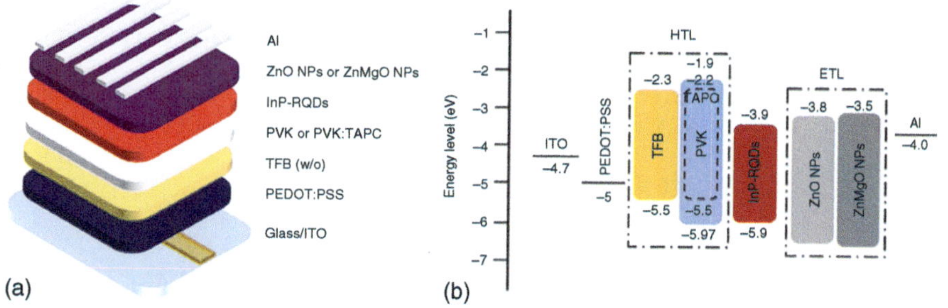

Figure 7.8 (a) QD-LED planar structure. (b) Energy band diagram of varied HTLs and ETLs. Source: Zhu et al. [18], with permission of Elsevier.

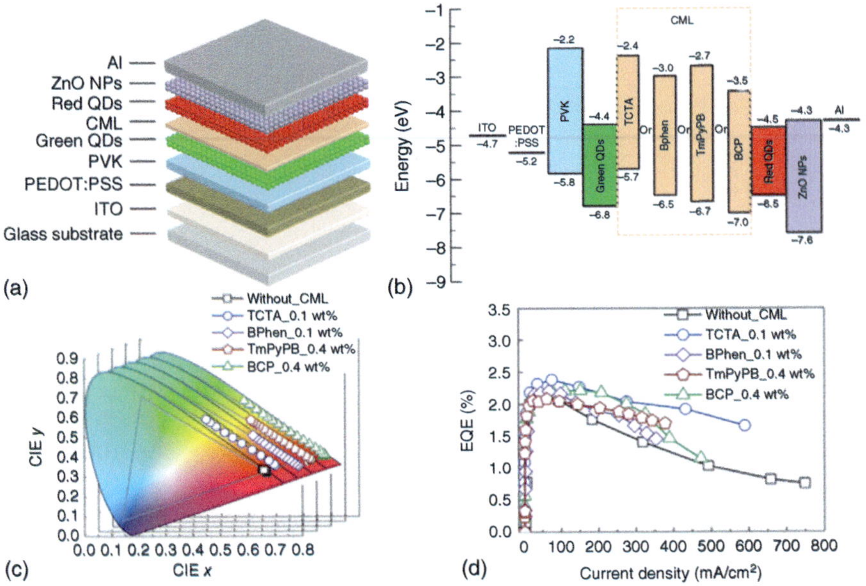

Figure 7.9 (a) QD-LED device structure. (b) Various CML candidates and their energy level diagrams. (c) Comparisons of double EML QDs prepared with different interlayer CMLs. (d) EQE comparisons of different CMLs. Source: Park et al. [19], John Wiley & Sons.

Meanwhile, the ETL doped with MgO is able to prolong the exciton decay time, which results in slower electron injection. These phenomena result in carrier balance within the QD-LED and contribute to its better performance of 7.58 cd/A of current efficiency at 2.4 V [18].

In some advanced QD-LEDs, there were two different QDs stacked in multilayer, which could be used as a means of color tuning. In order to achieve this, a charge modulation layer (CML) inserted in between two different QDs, as shown in Figure 7.9a, was developed, and the variation of CML in thickness and material determines the output and quantum efficiency as shown in Figure 7.9b [19]. The CML is not only used to improve the color tuning range but also helps to balance the electron-hole injection, which eventually leads to balanced carrier concentrations as in Figure 7.9c. With all the CMLs proposed, the TCTA CML showed the best EQE of 2.4%, as shown in Figure 7.9d, and BCP CML resulted in maximum color range of 0.3405. This research also opens a pathway for simplification of fabrication process while improving its efficiency and wide color range.

7.5 QD-LED Fabrication

One of main challenges now is to bring the lab-scale QD-LED fabrication to large-scale manufacturing for commercialization in consumer and industrial use. Here, all the stability and high efficiency must be transferable, and above all, the material used must be heavy metal-free. The simplest way to deposit the QDs and

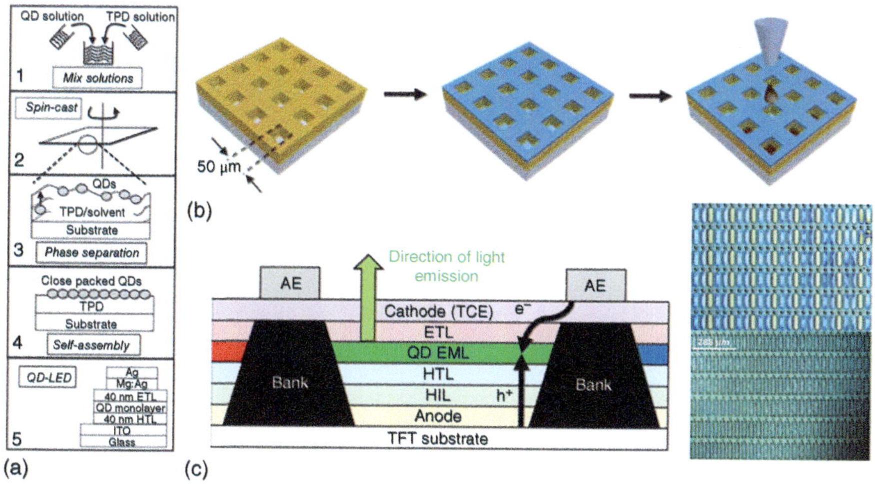

Figure 7.10 (a) Spin-coat assisted QDs formation by phase separation. (b) QDs ink jetted into micropatterns defined by lithography. (c) QD-LED pixelation with contact grid lines (left), device appearance before and after metal grid liners. Source: Bulović and coworkers [21, 22]; Hopkin et al. [23]/with permission of John Wiley & Sons.

the associated ETLs and HTLs can be done via spin coating process. The spin coating process is deemed to be compatible with large-scale manufacturing, where the QDs thickness can be controlled at nanoscale in order to achieve high luminance and EQE [20]. One may need to adjust the spinning speed according to the QDs material concentration and depending on the underneath material surface roughness. This can be challenging in case to get uniform thickness across the large substrates, where the edges of substrates tend to have much thinner thickness compared to center point of QDs dispensed. This spin coat method was further improved by the phase separation during coating process and self-assembly at subsequent step, where it is shown that a large area can be deposited with QD monolayer uniformly, as shown in Figure 7.10a [21]. A more innovative way was to use the photolithography method, as shown in Figure 7.10b [22]. It was demonstrated that the photoresist is first patterned on the electrode in pixels, then the ETL is spin-coated. Later, the QDs were ink jetted into the patterned pixels. This unique method can be adopted to use Cd-free QDs, and multi-QDs can be deposited into pixels to obtain multicolor red/green/blue emissions. Meanwhile, pixelation also requires delicate contact electrodes since the pitch distances are in sub-microns. In a recent study, a solution-processed transparent electrode, ITO, and Al metal liners were used to reduce the sheet resistances and increase ITO transparency as shown in Figure 7.10c [23]. This new method could be useful for the common front-side light-emitting QD-LED.

The photolithography method was taken in further experiment, which directly patterned the QDs using ultraviolet exposure. The study exhibits promising future large-scale fabrication since a uniform pixel was produced with room for improvement in terms of luminance and EQE, as shown in Figure 7.11a,c [24]. Figure 7.11b

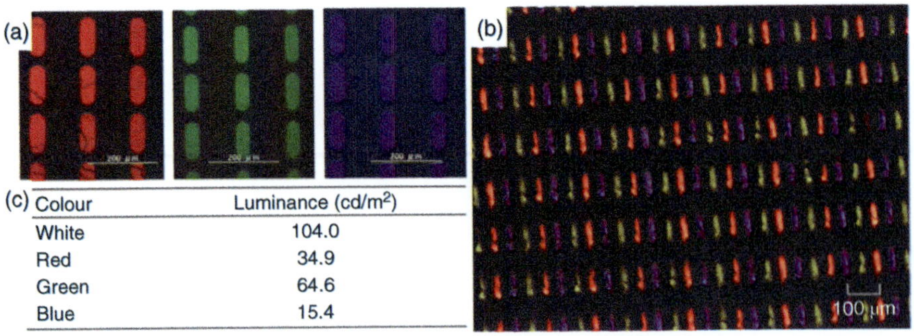

(c)

Colour	Luminance (cd/m²)
White	104.0
Red	34.9
Green	64.6
Blue	15.4

Figure 7.11 (a) UV patterned red, green, and blue QDs. (b) A matrix of red, blue, and green. (c) Corresponding luminance of QDs. Source: Nakanishi et al. [24]/with permission of John Wiley & Sons.

illustrates the alternating red, green, and blue emission in 6.24 inch QD-LED display.

The use of lithography technology, able to produce accurate pixels and high resolution, is achievable as this is compatible with the existing tools. We have seen the progress of QDs in their synthesis, device structures, and finally their possible integration in large-scale fabrication facilities. With extensive research in core-shell-surface ligand, the QY has been pushed well above 95% and narrow emission of 20 nm has been made possible. Furthermore, the Cd-free QDs also motivate the commercialization of high-brightness displays and smart and high-resolution displays.

References

1 Han, H.-V., Lin, H.-Y., Lin, C.-C. et al. (2015). Resonant-enhanced full-color emission of quantum-dot-based micro LED display technology. *Optics Express* 23 (25): 32504.

2 Bang, S.Y., Suh, Y.H., Fan, X.B. et al. (2021). Technology progress on quantum dot light-emitting diodes for next-generation displays. *Nanoscale Horizons* 6 (2): 68–77.

3 Bozyigit, D. and Wood, V. (2013). Challenges and solutions for high-efficiency quantum dot-based LEDs. *MRS Bulletin* 38 (9): 731–736.

4 Manders, J.R., Qian, L., Titov, A. et al. (2015). Next generation display technology: quantum dot LEDs. In: *SID Symposium Digest of Technical Papers*, vol. 46, 73–75. Wiley.

5 Jose, R. and Al-Douri, Y. (2022). Introduction to quantum dots. In: *Graphene, Nanotubes and Quantum Dots-Based Nanotechnology* (ed. Y. Al-Douri), 579–599. Elsevier.

6 Gaponenko, S.V. (1998). *Optical Properties of Semiconductor Nanocrystals*. Cambridge University Press.

7 Jose, R., Zhanpeisov, N.U., Fukumura, H. et al. (2006). Structure–property correlation of CdSe clusters using experimental results and first-principles DFT calculations. *Journal of the American Chemical Society* 128 (2): 629–636.

8 Murray, C.B., Norris, D.J., and Bawendi, M.G. (1993). Synthesis and characterization of nearly monodisperse CdE (E = sulfur, selenium, tellurium) semiconductor nanocrystallites. *Journal of the American Chemical Society* 115 (19): 8706–8715.

9 Coe-Sullivan, S., Liu, W., Allen, P., and Steckel, J.S. (2013). Quantum dots for LED downconversion in display applications. *ECS Journal of Solid State Science and Technology* 2 (2): R3026–R3030.

10 Bang, S.Y., Fan, X.B., Jung, S.M. et al. (2020). Highly stable and scalable blue QD-LED via an evaporated TiO_2 thin film as an electron transport layer. *Advanced Optical Materials* 8 (21): 2001172.

11 Kim, K., Suh, Y.H., Kim, D. et al. (2020). Zinc oxo clusters improve the optoelectronic properties on indium phosphide quantum dots. *Chemistry of Materials* 32 (7): 2795–2802.

12 Bai, J., Hou, T., Zhang, M. et al. (2022). An alloyed mid-shell strategy assisted realization of thick-shell violet-blue ZnSe/$ZnSe_xS_{1-x}$/ZnS quantum dots with high color purity. *Journal of Luminescence* 252: 119391.

13 Ippen, C., Guo, W., Zehnder, D. et al. (2019). High efficiency heavy metal free QD-LEDs for next generation displays. *Journal of the Society for Information Display* 27 (6): 338–346.

14 Kim, H.Y., Park, Y.J., Kim, J. et al. (2016). Transparent InP quantum dot light-emitting diodes with ZrO_2 electron transport layer and indium zinc oxide top electrode. *Advanced Functional Materials* 26 (20): 3454–3461.

15 Won, Y.H., Cho, O., Kim, T. et al. (2019). Highly efficient and stable InP/ZnSe/ZnS quantum dot light-emitting diodes. *Nature* 575 (7784): 634–638.

16 Kirkwood, N., Singh, B., and Mulvaney, P. (2016). Enhancing quantum dot LED efficiency by tuning electron mobility in the ZnO electron transport layer. *Advanced Materials Interfaces* 3 (22): 1600868.

17 Alexandrov, A., Zvaigzne, M., Lypenko, D. et al. (2020). Al-, Ga-, Mg-, or Li-doped zinc oxide nanoparticles as electron transport layers for quantum dot light-emitting diodes. *Scientific Reports* 10 (1): 7496.

18 Zhu, X., Liu, Y., Liu, H. et al. (2021). Optimization of carrier transport layer: a simple but effective approach toward achieving high efficiency all-solution processed InP quantum dot light emitting diodes. *Organic Electronics* 96: 106256.

19 Park, S.J., Song, S.H., Kim, S.S., and Song, J.K. (2021). Charge modulation layer and wide-color tunability in a QD-LED with multiemission layers. *Small* 17 (17): 2007397.

20 Bae, W.K., Lim, J., Lee, D. et al. (2014). R/G/B/natural white light thin colloidal quantum dot-based light-emitting devices. *Advanced Materials* 26 (37): 6387–6393.

21 Coe-Sullivan, S., Steckel, J.S., Woo, W.K. et al. (2005). Large-area ordered quantum-dot monolayers via phase separation during spin-casting. *Advanced Functional Materials* 15 (7): 1117–1124.

22 Azzellino, G., Freyria, F.S., Nasilowski, M. et al. (2019). Micron-scale patterning of high quantum yield quantum dot LEDs. *Advanced Materials Technologies* 4 (7): 1800727.

23 Hopkin, H.T., Boardman, E.A., and Smeeton, T.M. (2020). Solution-processed transparent top electrode for QD-LED. In: *SID Symposium Digest of Technical Papers*, vol. 51, 516–519. Wiley.

24 Nakanishi, Y., Takeshita, T., Qu, Y. et al. (2020). Active matrix QD-LED with top emission structure by UV lithography for RGB patterning. *Journal of the Society for Information Display* 28 (6): 499–508.

8

Ultraviolet LED Packaging and Application

The development of semiconductor light sources has achieved great milestones, from infrared to visible wavelengths. However, in the ultraviolet (UV) region, they have yet to reach their full potential. Advancements have been made in material growth, design, fabrication, characterization, and understanding of their device physics, but several difficulties related to insufficient light extraction, high dislocation density, and contact resistance still need to be elucidated [1]. Even with these problems, the current generation of commercial UV LED devices has impressive performance, long lifetime, and output power sufficient for many applications in comparison to standard lamp-based UV light sources [2]. This sparks great interest in UV application industry to replace their conventional UV light sources with UV LEDs.

8.1 UV LED Application

As illustrated in Figure 8.1, the UV applications are spread into three main categories, namely UV-A, UV-B, and UV-C. These three types of UV are defined by their wavelengths. Based on the convention established during the Second International Congress on Light in 1932, the UV spectrum is classified as UV-A (315–390 nm), UV-B (280–315 nm), and UV-C (200–280 nm). The boundary between UV and visible is at about 390 nm, where the 1987 CIE eye sensitivity curve has a value of 0.1% of the maximum value. Any wavelength from 390 to 740 nm is considered as visible light [3].

The UV applications in Figure 8.1 give a general landscape of the application areas of the UV LED. The market is predominantly based on the replacement of traditional UV lamps. However, in the meantime, this opens new market that traditional UV lamps are unable to access. This is mainly due to compactness, longer lifespan, and robustness of UV LEDs that perfectly fit handheld application devices. The UV LED performance over the year has been improving drastically [4].

The application technology for UV-A as illustrated in Figure 8.2 is mainly for UV curing process for materials that require UV light for curing. UV-A is also used for counterfeit detection and photocatalytic purification. In the consumer industry, it is commonly used for tanning purposes for those who stay far from the equator

LED Packaging Technologies: Design, Manufacture, and Applications, First Edition.
Luruthudass Annaniah, Mohamed Salleh M. Saheed, and Rajan Jose.

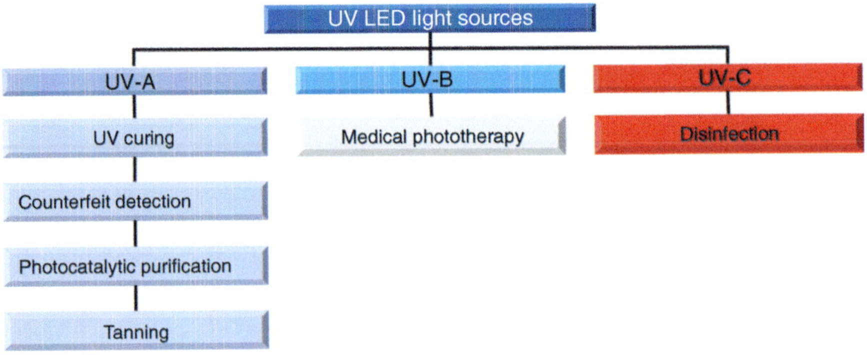

Figure 8.1 Shows categories of UV light sources and their application.

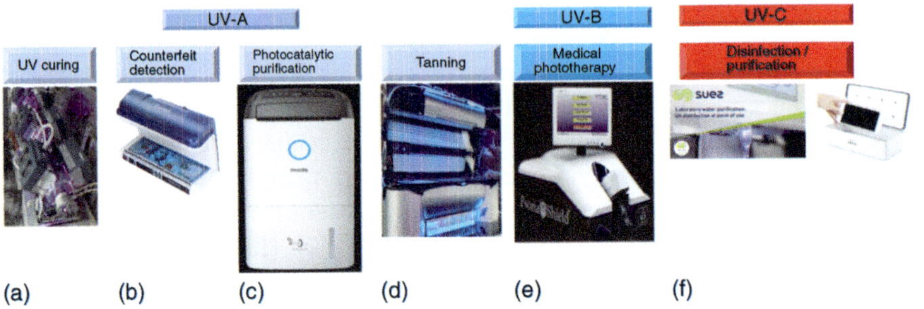

Figure 8.2 UV-A, UV-B LED, and UV-C ranges of applications. (a) Semray@ UV PC6003, a UV LED curing system. Source: Courtesy of Heraeus Nobleight GmBH. (b) TIHOO T0020 counterfeit money detector. Source: TIHOO Technology. (c) Philip DE5205. Source: Koninklijke Philips N.V. (d) Tron Sunbed. Source: Smart Sunbeds UK. (e) Skin phototherapy. Source: Psoria-Shield, Inc. (f) Water Purification. Source: SUEZ. (g) Phone Disinfection Box. Source: Shenzhen UVLED Optical Technology Co. Ltd.

where there is lack of sunlight. UV-B, on the other hand, is mostly used for medical phototherapy, i.e. skin treatment. The main driver in this area is Psoria-Shield, a new player in UV LED phototherapy player. UV-C, the most powerful high-energy wavelength UV light source, is among the three categories mostly used for disinfection [5]. It kills germs. White goods makers like LG have commercialized UV-C LEDs in their fridges [2]. In addition, handphone makers also started research using UV-C disinfection phone box, as illustrated in Figure 8.2.

UV application was first introduced with mercury vapor lamps. It has its downsides in comparison to LED-based UV. The advantages and disadvantages are explained in Table 8.1. The vapor lamps are large, on average, at 1 m compared to UV LED of 1 cm^2. UV lamps consist of toxic mercury, while LEDs have no toxic elements. The lifetime of UV lamps is in the range of 1000–20,000 hours. UV LEDs are highly dependent on the wavelength, and with current technology, the UV-A lifetime is in the range of 10,000–50,000 hours [6]. UV-B and UV-C in the range of 1000–10,000 hours. There is no warm-up time needed for UV LED compared to UV lamps, which take about 10 min. The robustness of UV LEDs is considerably better

Table 8.1 The advantages and disadvantages of UV LEDs and UV lamps.

	UV LED	UV lamp
Advantages	- UV LED is smaller compared to UV lamps Lifetime 10,000–50,000 hr - Spontaneous switch on. No warm-up is needed. Robust package - UV-A wall-plug efficiency is higher than UV lamp - Overall cost of ownership is cheaper than UV lamp	- UV-B and C lamps have better wall-plug efficiency compared to UV-LED
Disadvantages	- UV-B and C wall-plug efficiency is lower than UV lamp	- UV lamp is very large (1 m length) compared to UV LED (1 cm^2) - Lifetime short – 1000–20,000 hr - Warm-up time is roughly 10 min - Fragile-made of quartz that are breakable - Not design friendly – Lamp design must be straight and long - Overall cost of ownership is higher than UV LED

in terms of shock-resistance compared to UV lamps made up of quartz, which are breakable. The design of UV LEDs can be perfectly adjusted to the system due to their small size. UV lamp design must be straight and long. The overall cost of the UV LED is relatively low compared to the UV lamp. The other advantages of UV LEDs come from their long lifetime and low power consumption [2]. UV-A LED wall-plug efficiency is higher than UV-A lamp. However, in the UV-B and UV-C categories, the UV lamp wall-plug efficiency is greater than the UV LED.

The UV LED market is expected to reach USD 650 million in the year 2021[1]. This is driven by replacement of UV lamps with LED and a paradigm shift in the visible LED market due to intense cost pressure in conventional visible LED market that focuses on UV LED, where price is higher compared to conventional visible LED. Many LED makers are now focusing on UV LED to extend their product portfolio for a better margin. To name a few, there are Nichia, Panasonic, Lumiled, LG, Samsung, OSRAM, Everlight, CREE, Seoul Semiconductor, Sanan, and many more new entries.

UV LED industry is not as rosy as it has its challenges. UV-A is close to visible range wavelength, making switching from blue LED to UV-A easier as current blue LED packaging technology is easily adaptable to UV-A. However, UV-B and C wavelengths are shorter than visible LED. This high energy electromagnetic wavelength spectrum harms organic base encapsulant materials. The carbon compound in the packaging material degrades as a result. This process is called carbonization. The efficiency of the chip is also rather low for UV-C. As a result, the energy injected is mostly converted to heat. This heat must be channeled out of the chip through proper packaging design. Hence, thermal management is very crucial in package design.

Even though, with all these challenges in UV LED, the reward is more than risk. Hence, many players are putting in a lot of effort to make UV LEDs. In the next subchapters, the packaging technology for UV LEDs will be elaborated.

8.2 UV-A and B LED Packaging Technology

UV-A is generally not harmful and can be used for many applications. As illustrated in Figure 8.2, one of the applications is in tanning. It is generating a lot of interest, especially in countries that have less sunlight exposure. The wavelength that covers tanning is from 305 to 370 nm. The output power for this application varies from 0.5 to 10 W/cm^2. Ideally, for this wide wavelength range and power, an array of metal-can, aluminum packages or ceramic packages assembled in one system is a perfect fit. Here, the system requires excellent external heatsink to dissipate the heat from the package [7].

Figure 8.3 illustrates several types of UV-LED packages that cover UV-A to UV-C with different powers. The higher the power, the more the package requires efficient thermal management package design [8]. In some cases, thermal via has to be incorporated with a heatsink to enable efficient heat dissipation, as illustrated in Figure 8.4. UV-B and UV-C chips in principle have lower IQE, hence more energy is converted to heat. Hence, a ceramic or aluminum package with a glass cap coupled with good heat dissipation package design fits well with this package [7]. UV-C, in particular, requires special glass cap that is made of hardened glass, as the UV-C wavelength is much more corrosive.

The other application of UV-A LED is for curing processes. This is concentrated on curing systems for inks, coatings, and adhesives. This application requires high-power UV LED. They are in the range of 1–10 W/cm^2. The package used

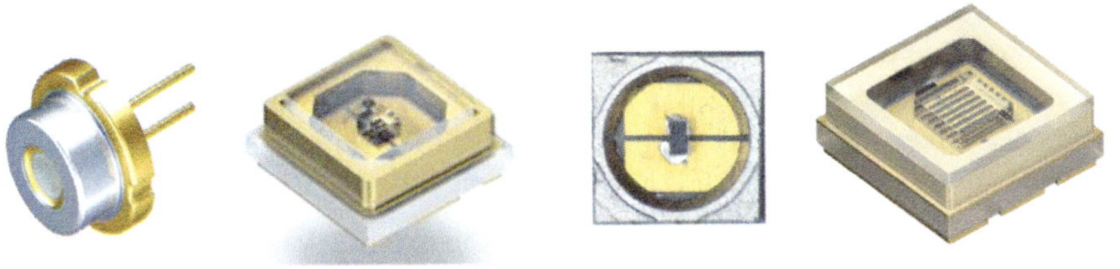

(a) (b) (c) (d)

Figure 8.3 UV-LED packages for various power – Metal Can, ceramic, and aluminum packages [8]. (a) Metal Can package TO 90 for UV-A and B low to mid power application. Source: Courtesy of ams OSRAM GmbH. (b) Ceramic package for UV-C Mid power application. Source: Courtesy of ams OSRAM GmbH. (c) Aluminum package for UV-B and C low- to mid-power applications. Source: Pai et al. [8]/MDPI/CC BY 4.0. (d) Ceramic package with glass cap for UV-B and C mid- to high-power applications. Source: Courtesy of ams OSRAM GmbH.

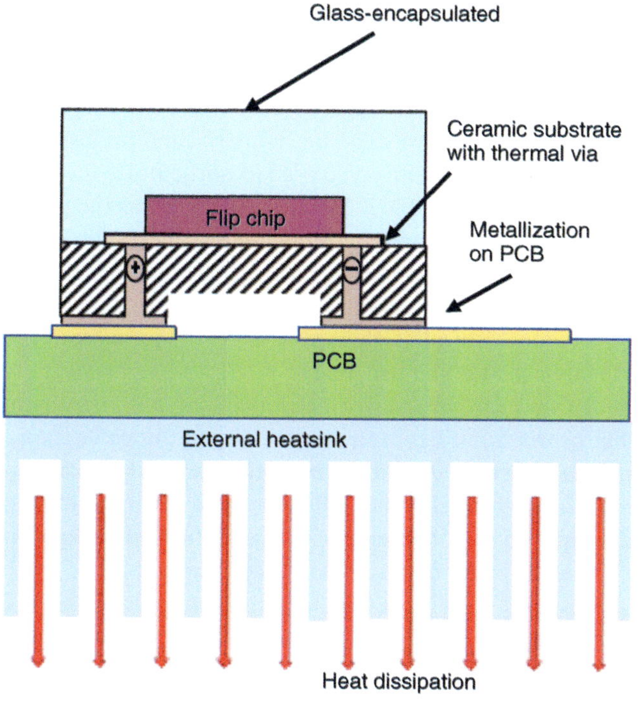

Figure 8.4 Cross-section view of ceramic package with glass cap UV soldered on a luminaire PCB with external heatsink.

either aluminum or ceramic and was sealed with a glass cap. Due to the high-power application, thermal management is very critical for this application. Usually, the ceramic package has thermal via to connect to an external heatsink for dissipating the excess heat generated from the product. This is shown in Figure 8.4. The UV chip is soldered on to the ceramic substrate, and the thermal via is directly connected to external contact that will be soldered to an external heatsink. This ceramic package with a glass cap can also be used for UV-B and UV-C mid- to high-power applications.

In UV curing applications, the UV LED product's design is based on several parameters, and each will impact the overall product performance. The LED choice for curing applications is influenced by the package design, the thermal management, the optics, and power supply.

UV-A is also used as a light source for counterfeit detection for banknotes, passports, fake goods, and others. They have used a range of areas, right from banks to casinos, immigration, policing, and securities. The market drivers for this application mostly focus on low cost, compactness, high lifetimes, and low power consumption. The wavelength that fits this application is in the range of 365–385 nm [6]. This is a low-power LED that is less than 0.1 W/cm^2. Hence, mostly use Metal Can package, as illustrated in Figure 8.3, which is suitable for low-power LEDs and a low-cost applications.

UV-A LED is also used for photocatalytic purification. This is mainly in an air conditioners and standalone air purifiers. Air purification is UV LED's biggest potential because the photocatalysis is active against a large spectrum of air pollutants (microbes, mold, odors, etc.) [9]. Small air purifiers for consumers are mostly for automotive and residential use. There are also some for industrial and professional use, primarily in hospitals. UV-A LED is recently also used in horticulture industry. For this application, the package that is suitable is Metal Can package as the power requirement is close to 1 W/cm^2. Epoxy-based packages are not suitable as the power requirement is rather high. The encapsulant in Metal Can package is metal and glass. In some cases, the ceramic package with a glass cap was used.

UV-B LED is mostly used for medical phototherapy and curing processes. The medical phototherapy is mostly used for skin treatment. Studies show exposure to UV-B light, particularly in 310 nm narrowband, is an effective long-term treatment for many skin conditions like psoriasis, vitiligo, and eczema. The output power range varies depending on the treatment required. The market drivers are focused on narrow UV-B spectrum, compactness, long lifetime, efficiency, and precise emission sources [10]. For this type of application, the best package is Metal Can package or ceramic package, arranged in an array for this high-power application.

8.3 UV-C Packaging Technology

The UV-C LED application mostly falls into the disinfection market [11]. In early 2020, the World Health Organization (WHO) officially announced a health warning and declared COVID-19 pandemic. This situation creates a boom in disinfection industry. Hospitals, quarantine quarters, workplaces, and public amenities all need quick disinfection to avoid spreading this deadly pandemic virus. UV-C is the ideal wavelength for disinfection as the wavelength range of 253.7 nm has the highest absorption rate at the virus DNA based on studies by Memarzadeh, Olmsted, et al. Their studies in healthcare facilities show that UV-C light is microbiocidal. The airborne microorganisms can be killed or inactivate infectious microorganisms once they are exposed to UV-C light with the right UV-C power intensity and exposure time. In the same report, Mermarzadeh, also mentioned that biological effects in humans from overexposure to UV-C radiation vary with wavelength, photon energy, and duration of the exposure [12].

The science of UV-C disinfection is that when the microorganism such as bacteria, viruses, fungi, and simple life forms are exposed to UV-C light, the light penetrates through their cell walls and disrupts the structure of their DNA, thereby, prohibiting reproduction [13]. Once it is unable to reproduce, the ability to grow and multiply is also stopped, and it is classified as cellular death, and it is harmless and no longer pathogenic [12, 14].

The biggest challenge in UV-C application is overall system efficiency [2]. The chip efficiency at present moment is rather low compared to UV-C lamp-based applications. Hence, adoption is still slow. Moving forward, as the chip's efficiency improves, the application potential and range are also expected to increase. UV-C is considered high-energy, short-wavelength spectrum that is more corrosive compared to UV-B or A. Hence, for this UV wavelength category, with the current low-efficiency chips, the package has to be ceramic or metal-based with a glass cap. This package, as illustrated in Figure 8.3, is suitable as it has desirable heat dissipation capabilities and is stable for UV-C application. Some of the main suppliers of UV-C LEDs are OSRAM, Nichia, Luminus, and LG.

8.4 Future Application of UV-LED and Packaging Design Evolution

The new applications for UV-LED in the mid- and long-term are mainly in dialysis, water disinfection, purification, surface disinfection, dishwashers, portable hand disinfection, contact lens boxes integrating with UV LED, and many more. The UV-C LEDs have incredible opportunities in future. However, it has to overcome the key challenges that a robust and compact package has to ensure the lifetime and reliable performance to meet the market demands. Even though the current efficiency of UV-C LED is not as high compared to that of standard UV lamps, it is steadily improving. Since the efficiency is low, the package design plays a very important role to ensure no photons are lost and the thermal management is very efficient. Many research projects are underway to fill the gap in UV-LED package. Some of the package options are through-silicon-via (TSV) package [15], liquid packing structure (LPS), and surface-mount aluminum package with a glass cap. The next subsection reviews LPS packaging.

8.4.1 Novel Liquid Packaging Structure

It is crucial and of the utmost importance to realize high-efficiency UV-LED, by enhancing their light-extraction efficiency and reducing thermal resistance. In 2019, Kang, Lin, et al. reported that a liquid packaging structure, as illustrated in Figure 8.5, could enhance optical power by 27.2% and 70.7% for flat-type and lens-type UV-LEDs, respectively [16].

The liquid that was used as an encapsulant is a silicone oil that covers the UV chip. It seems the combined silicone liquid and ceramic package structure reduces the thermal resistance significantly by 30.3% compared to the conventional structure, as the study claimed. In the reliability test up to 200 hours, the package showed good thermal and deep UV resistance at a wavelength of 268 and 310 nm, based on their observations. Their work also shows the reliability of liquid packaging with DUV-LEDs compared to the conventional structure.

This packaging structure is feasible as a compact structure with high light extraction efficiency and promising good thermal management of deep UV LED packaging, and it may inspire others to follow in near future as this packaging concept as it is relatively less complicated to manufacture compared to other packages.

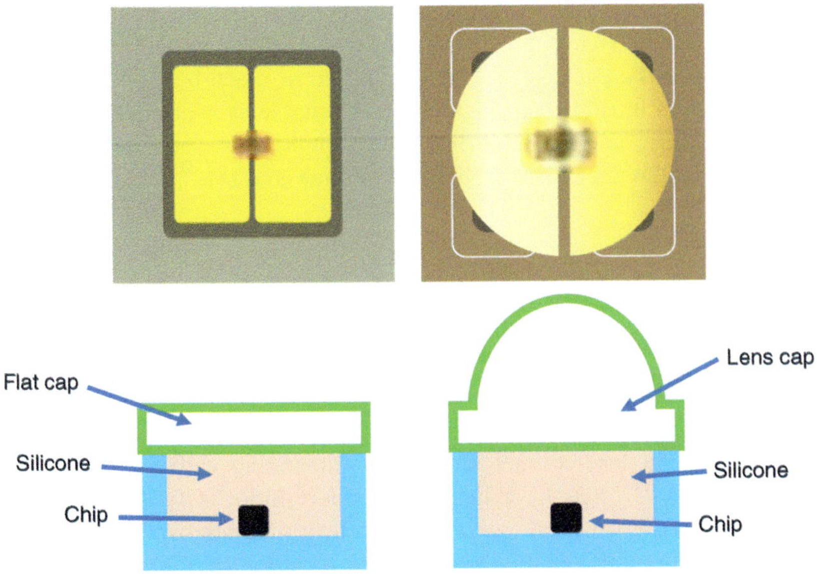

Figure 8.5 Liquid package structure for Deep UV LED. Source: Kang et.al. [16] 2019/MDPI/CC BY 4.0.

8.5 Impact of UV-LED to UV Light Source Business

The technology transition to UV LEDs is one of the greatest effects on pure UV lamps manufacturers over the next few years. UV LED's higher device lifetimes and low cost will directly impact the lamp replacement market and will shift these values to the system level in mid- and long-term [17]. For traditional UV lamp manufacturers, there is a risk of losing revenue with UV LEDs' emergence. However, for the manufacturers of UV system, the transition to this new technology is simpler, since they are integrators of UV light sources. It will be easier to enter the UV system market based on their existing know-how. Moreover, they already won distribution channels. Pure UV lamp manufacturers are another parallel story. They will need to develop new competencies in order to access UV LEDs. The best option they have consisted of manufacturing UV LEDs modules/arrays. However, as value consistently moves into the system, the best choice for most of them will be to diversify their activities and enter the manufacturing of UV LED systems to obtain additional revenue.

8.6 Summary

The LED has evolved toward UV wavelengths. The applications in this wavelength category are getting much attention, especially for disinfection purposes after the world experienced COVID-19 pandemic. Besides this, there are also many other novel applications of UV LED that are still not fully tapped. The packaging for UV

LED is slightly different as the UV wavelength is considered high-energy wavelength where it damages the molecules of the carbon-based encapsulant. Hence, mostly for UV LED packaging, ceramic or metal with glass cap-based package were used. However, more research is ongoing to further improve the UV LED package and efficacy. The future of UV LED is very bright!

References

1 Taghipour, F. (2018). UV LED technology: the times they are a-changin'. *IUVA News* 20 (1): 14–17.

2 Muramoto, Y., Kimura, M., and Nouda, S. (2014). Development and future of ultraviolet light-emitting diodes: UV-LED will replace the UV lamp. *Semiconductor Science and Technology* 29 (8): 084004.

3 Schubert, F.E. (2006). *Light Emitting Diode*, vol. 2. New York, USA: Cambridge University Press.

4 Huang, J.J., Kuo, H.-C., and Shen, S.-C. (2014). *Nitride Semiconductor Light-emitting Diodes (LEDs) Materials, Technology and Applications*. Cambridge, UK: Woodhead Publishing.

5 Anderson, M. (2020). The ultraviolet offense: Germicidal UV lamps destroy vicious viruses. New tech might put them many more places without harming humans. *IEEE Spectrum* 57 (10): 50–55.

6 Halliday, R. (2014). *Key benefits of next-gen UV LED technology*. Lumex®(4 pages) Downloaded from website on Aug, 2014. 21.

7 Peng, Y., Liang, R., Mou, Y. et al. (2019). Progress and perspective of near-ultraviolet and deep-ultraviolet light-emitting diode packaging technologies. *Journal of Electronic Packaging* 141 (4).

8 Pai, Y.-M., Lin, C.-H., Lee, C.-F. et al. (2018). Enhancing the light-extraction efficiency of AlGaN-based deep-ultraviolet light-emitting diodes by optimizing the diameter and tilt of the aluminum sidewall. *Crystals* 8 (11): 420.

9 OSRAM (2020). *OSRAM AirZing Mini - Purifies the Air So You Can Breathe Freely*. OSRAM. https://www.osram.com/am/automotive-care-and-equipment/airzing-mini.jsp.

10 Argyraki, A., Petersen, P.M., and Dam-Hansen, C. (2017). *New Light Sources for Biomedical Applications*. Technical University of Denmark (DTU).

11 Vitzilaiou, E., Kuria, A.M., Siegumfeldt, H. et al. (2021). The impact of bacterial cell aggregation on UV inactivation kinetics. *Water Research* 204: 117593.

12 Memarzadeh, F., Olmsted, R.N., and Bartley, J.M. (2010). Applications of ultraviolet germicidal irradiation disinfection in health care facilities: effective adjunct, but not stand-alone technology. *American Journal of Infection Control* 38 (5): S13–S24.

13 Song, K., Mohseni, M., and Taghipour, F. (2016). Application of ultraviolet light-emitting diodes (UV-LEDs) for water disinfection: a review. *Water research* 94: 341–349.

14 Martínez de Alba, A.E., Rubio, M.B., Morán-Diez, M.E. et al. (2021). Microbiological evaluation of the disinfecting potential of UV-C and UV-C plus ozone generating robots. *Microorganisms* 9 (1): 172.

15 Chiba, H., Suzuki, Y., Yasuda, Y. et al. (2022). DUV-LED packaging using high density TSV in silicon cavity and laser-glass-frit-bonded UV transmitting glass cap. *Sensors and Actuators A: Physical* 344: 113700.

16 Kang, C.-Y., Lin, C.-H., Wu, T. et al. (2019). A novel liquid packaging structure of deep-ultraviolet light-emitting diodes to enhance the light-extraction efficiency. *Crystals* 9 (4): 203.

17 Lawal, O., Cosman, J., and Pagan, J. (2018). UV-C LED devices and systems: current and future state. *IUVA News* 20 (1): 22–28.

9

Lifecycle Analysis and Circular Economy of LEDs

9.1 Introduction

LEDs are currently the lighting device of choice for low carbon footprints – create only ~200 kg of CO_2 per year. A study by the Environmental Protection Agency found that by replacing only one light bulb in every house in the United States with an LED bulb would reduce greenhouse gas emissions by ~4 million metric tons [1–3]. Lighting accounts ~5% of the total greenhouse gas emissions; and a global switch to LED technology can have significant impact on environment, i.e. saving over 1400 million tons of CO_2 per year and avoiding the construction of 1250 power stations [1–3]. However, the total energy involved and environmental impacts of any product, including LEDs, are to be accurately assessed over its entire lifecycle, i.e. production, transportation, consumption, and end-use management, in order to scrutinize a technology for sustainability. This process is called the lifecycle analysis or LCA or is often described as cradle-to-grave analysis, as shown schematically in Figure 9.1.

Circularity is one of the fundamental characteristics of LCA and is the natural algorithm for sustainability. On the other hand, a linear approach of making products, using them, and discarding them at their service life has been practiced since the industrial revolution. This approach of make, use, and discard is known as linear economy; the linear economic model has been widely cited as the primary contributor to many of the existing existential threats, including environmental adversity, biodiversity loss, and resource scarcity. Realizing this, a circular economy is now widely discussed, which is about keeping products and resources at the highest quality and value for as long as possible, contributing to the restoration and regeneration of the environment, and addressing the supply chain management via a lesser dependence on the finite natural resources. For implementing circular economy, resources are expected to be part of regenerative and low-carbon processes. Figure 9.2 schematically illustrates circular economic approach.

As can be seen from Figure 9.1, the linear economic model generates enormous waste during each step of their lifetime by spending significantly high energy. On the other hand, waste is minimized or even eliminated in the circular economy through

LED Packaging Technologies: Design, Manufacture, and Applications, First Edition.
Luruthudass Annaniah, Mohamed Salleh M. Saheed, and Rajan Jose.

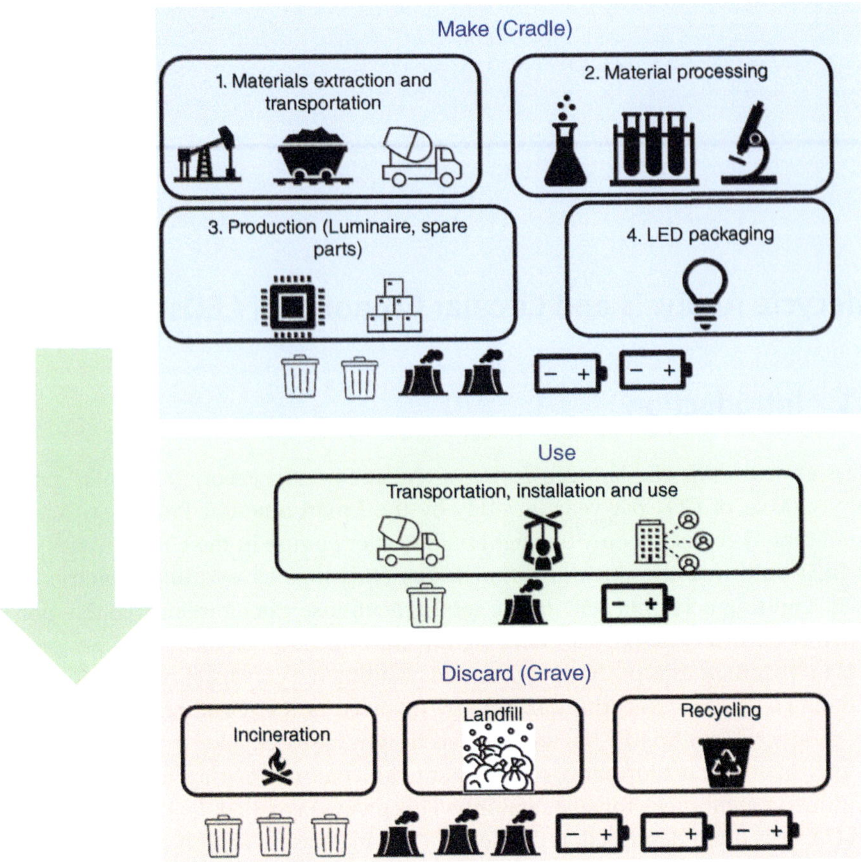

Figure 9.1 Linear economic processes of the LED lifecycle. The bottom panel of each domain illustrates the relative amount of waste, carbon footprint, and energy expenditure.

recycling, remanufacturing, reusing, or repurposing so that the dependence on limited natural resources can be minimized or eliminated. For these reasons, LCA is often re-termed from "cradle-to-grave analysis" to "cradle-to-cradle" analysis.

This chapter deals with the lifecycle analysis of LEDs and has been divided into LCA of LEDs and Circular Economy of LEDs.

9.2 LCA of LEDs

The LCA results are used to analyze (i) carbon footprint of materials in sourcing, transporting, and processing; (ii) the materials used for fabricating a device; (ii) energy spent; and (iv) waste generated in a technology and could be used to optimize the process for ideal usage times in terms of energy and material efficiency. As shown in Figure 9.1, the production of LEDs starts from materials sourcing through (i) mining of minerals for synthesizing inorganic materials used for fabrication of the luminaire, metallic and semiconducting interconnects, and their transportation to the sites where they are (ii) purified in addition to synthesis of polymers used and

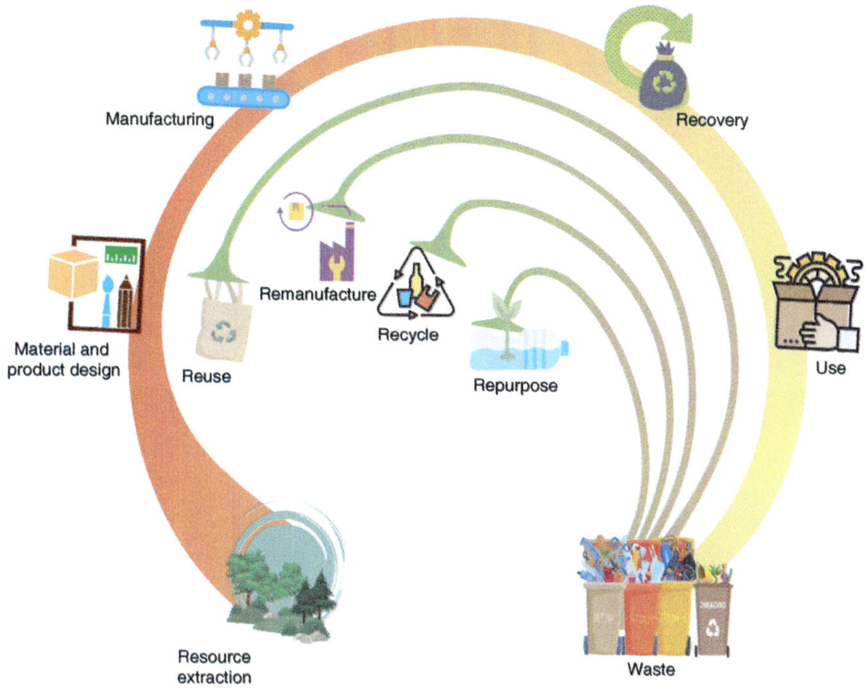

Figure 9.2 Schematics showing circular economy. The circular process starts with resource extraction, material and product design, manufacturing, use, and end-used products. After the use, some portion can be recovered, reused, repurposed, recycled, or remanufactured.

characterized. Refer to Chapters 1 and 3 regarding the components and device structure of an LED. The next step is to fabricate the p–n junction diodes and other spare parts for LEDs and finally packaging them to make a functional LED. As shown in Figure 9.1, significant amount of energy is spent during the making process, which contributes to the carbon footprint and produces materials-related waste (materials footprint). Two significant quantities involved in LCA are the materials footprint and embodied energy.

9.2.1 Materials Footprint

As explained in the previous chapters, an LED package includes luminaire, which is a p–n junction diode fabricated using a range of compound semiconductors for specific color emission or their assembly for white-light LEDs, as well as their substrates and packaging materials such as thermosetting materials such as silicone, siloxane; thermo-plastic (PMMA), metal oxide semiconductors (TiO_2), and metals for interconnects (copper, gold, silver, etc.). All these materials are either to be sourced from Earth or synthesized from raw materials and vary appreciably for different types of LEDs. The total mass of the materials for a single LED varies in the range of 83–290 g, with an average of ~175.1 g. A breakdown of the materials used in two typical LEDs with varied material choices, as considered by Hendrickson et al., is in Table 9.1. Figure 9.3 shows the cost distribution of LED production.

Table 9.1 Weight of various components used in a single LED.

	Choice 1			Choice 2		
Function	**Part name**	**Material**	**Weight (g)**	**Part name**	**Material**	**Weight (g)**
Optics	Lens	Glass	21.8	White border lens	Plastic	22.6
	Heat sink retention ring	Aluminum	9	Lexan stiffener	Plastic	11.4
Housing						
	Plastic con	Plastic	4	5 machine screws $440 \times 3/8$ PHP, 2 copper sheet metal screws, SMS No. $4 \times 1/4$ PHP	Copper	3.5
				Structural con	Plastic	53.2
				O ring	Rubber	0.5
	Array of 9 LEDs in 1 array, sensor, substrate, thermal grease		1.5	Array of 18 LEDs in 3 arrays		3
LED Module						
	Local heat sink	Copper	28.2	Local heat sink	Aluminum	
Heat sink	LED base heat sink	Aluminum	147.4	Heat sink assembly	Aluminum, Copper	120.9
Base assembly	Insulating compound	Fiber glass	0.3	Edison base assembly	Plastic	10.8
	Porcelain base	Porcelain	19.4	Plotting compound	Silicon	6.8
Driver	Resistors, transistors, inductors, capacitors, and diodes		6.1	Resistors, transistors, inductors, capacitors, and diodes	PCB, copper wire, and Teflon tubing	6.7
Edison Screw	Edison screw base	Tin-plated steel	9.1	Edison screw base	Tin-plated steel	8.6
Total weight			246.8			282.0

Source: Data adopted from Hendrickson et al. [4].

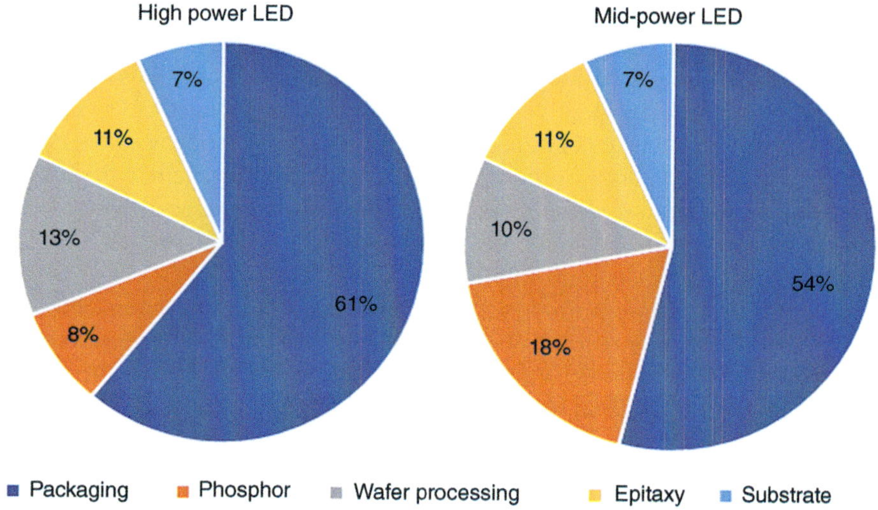

Figure 9.3 Cost division for high-power and mid-power LEDs. Data Source: LEDCOM model with inputs from DOE SSL Roundtable and Workshop attendees. Source: Adapted from LEDCOM model with inputs from DOE SSL Roundtable and Workshop attendees.

As detailed in Table 9.1, the primary functional component of an LED, i.e. the LED chip fabricated using compound semiconductors and phosphors on PCBs, has the lowest material usage of ~6.7–7.8% in both device choices. On the other hand, the metallic parts used for packagings, such as aluminum and copper, account for 50–60%. Plastics used for packaging claim the remaining material fraction. The large fraction of metals and plastics in the packaging offers a possibility of getting them recycled. Although it appears from Figure 9.3 that the primary cost factor of LEDs is packaging, the normalized cost per material is much higher for the LED chip and associated primary functional components. Given that several billionunits of LEDs are produced annually, the material footprint that the technology leaves is significant.

9.2.2 Embodied Energy and Carbon Footprint

Next major concern is the embodied energy, i.e. total energy expenditure for the production of an LED package, right from the material sourcing to the final destination of their application, including transportation and the associated carbon footprint. The embodied energy is usually represented by 20 million lumen-hours of output, which is equivalent to the service offered by a 60 W LED. The LED manufacturing requires much more energy than other lighting options (incandescent and CFLs). On an average, ~40% of the energy needed in LED manufacturing is spent on the fabrication of the chip, PCB, and circuit components, ~35% on packaging, and ~25% on the processing of bulk materials. While energy required for processing has been widely documented, those for sourcing (mining) are relatively underreported, primarily because of the uncertainties and larger errors it offers due to differences in the geographical locations, mining depths, and transportation. This limitation makes the embodied energy calculations underestimated. Besides, the environmental footprints associated with mining are often ignored in LCA studies. In the following, an approximate idea of the energy requirement to produce LEDs is given.

Among the materials listed in Table 9.1, all the inorganic components (aluminum, copper, steel, silicon, gallium, indium, etc.) are conventionally obtained from mining and subsequent processing, whereas the plastics are synthesized mostly from petrochemicals. Aluminum and gallium are extracted from the natural mineral bauxite; the energy expenditure to process 1 ton of alumina (Al_2O_3) is ~14.5 GJ. Aluminum is obtained from Al_2O_3 via smelting through the Hall–Heroult Process at ~940–980 °C, representing considerable energy expenditure during processing. Given that several billion units of LEDs are produced annually and that aluminum is the largest material fraction used in them, the energy expenditure for aluminum production is large. The separation and purification of the gallium and growth of single crystal gallium nitride substrate are much more energy intensive, and typically requires ~100 atm. of pressure and >750 °C of heat. Plastic production is also energy intensive. Khripko and colleagues report a value in the range of ~17.24–17.85 kWh per kilogram of polymer. LED chip fabrication is both energy- and material-intensive. The energy requirements arise from the metal-organic chemical vapor deposition (MOCVD) process, which is performed at high temperatures (400–1300 °C) for prolonged durations depending on the required thickness of the film. The MOCVD films are then to be cooled, requiring a large amount of cooling water. Besides, large amounts of gases such as hydrogen are required to carry the metal (gallium, indium, arsenic, etc.) vapor to the substrate. Significant amounts of energy are required in producing and purifying the hydrogen gas as well as the pumping the cold water through the system. The transportation contributes to every step of LED production; however, transport contribution to the embodied energy is typically >1% of the total lifecycle energy of LEDs.

The average manufacturing phase primary energy of a 2011 LED as compiled by the US Department of Energy for a module containing 16 LED packages is 343 MJ/20 million lumen-hours. The respective average contributions from materials and total LED package (including the chip fabrication) are 87 (~25%) and 256 MJ/20 million lumen-hours (~75%). The average manufacturing phase primary energy of incandescent (42.2 MJ/20 million lumen-hours) and compact fluorescent lamps (170 MJ/20 million lumen-hours) is significantly lower than that of LEDs. However, the considerably longer lifetime of LEDs and their lower energy requirements to power them than the other two luminaires efficiently offset the higher embodied energy of the former. The total carbon footprint of LEDs is 113 Kg CO_2Eq.

9.3 Circular Economy of LEDs

Due to the longevity of LEDs and their relative newness, their large-scale disposal and repurposing have not been well established. As shown in Figure 9.2, the circular economy is conventionally treated as end-use management of products, via

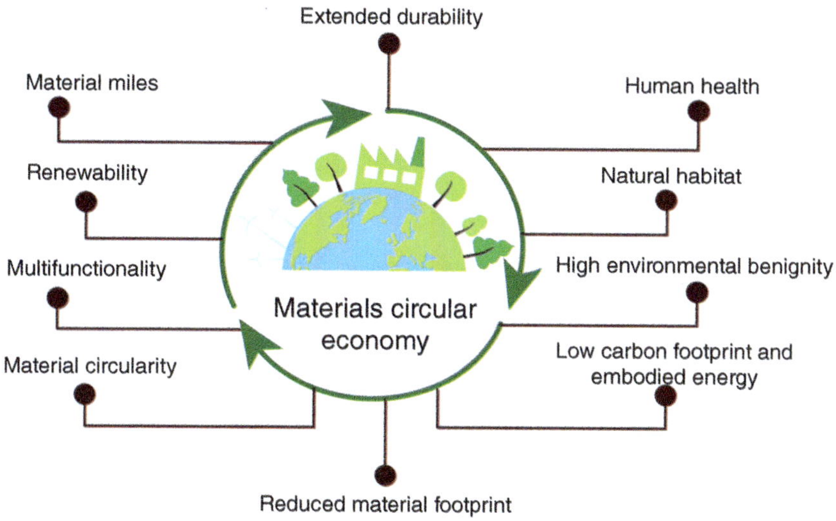

Figure 9.4 Schematic representation of 10 principles of materials circular economy. Source: Adapted from Ramakrishna and Jose [5].

recycling, recovering, reusing, repurposing, and remanufacturing. However, in a broader context, circular economy fosters a healthy living environment via sustainable use of materials by reducing the use of products or materials and elimination or reduction of resource wastage, resource depletion, and greenhouse gas emissions, which is termed as materials circular economy. In the material circular economy, the materials and products are purposely designed with lower environmental footprint and social costs and higher circularity while satisfying the cost as well as functional requirements. In this section, the end-of-use management of LEDs is addressed considering ten characteristics of materials' circular economy (Figure 9.4). These 10 strategies include (i) lower materials by design, (ii) multifunctional materials, (iii) materials of higher circularity, (iv) high durability materials, (v) low carbon and embodied energy materials, (vi) reduced material miles, (vii) sustainable materials, (viii) materials of higher environmental benignity, (ix) no toxicity, and (x) materials enabling natural habitat [5].

9.3.1 Lower Material Quantities by Design and Enhanced Material Properties

Selection, design, and development of materials to meet the required product functions and performance have always been the core pursuits of materials research. Products are often made with generous quantities of materials. This resulted in higher materials footprint per product and generation of large quantities of waste at the end-of-use of products. Efforts must be made to intentionally reduce the amount of material used in a product via enhanced material performance to reduce the materials intensity of a product and thus to reduce the waste generated at the end of product use. Hendrickson et al showed a strategy for a lower materials footprint,

Table 9.2 Weight of various components used in a single LED to reduce the materials footprint.

Part type	Name	Material	Mass (g)	Mass (%)
Optics	Glass bulb	Glass	10.7	13
LED Module	LED board connectors	Gold-plated copper	0.5	0.6
	9 LEDs in 1 array		1.5	1.8
Heat sink	Local heat sink ring	Aluminum	5.7	6.9
	Heat sink outer cone	Aluminum	18.1	22
	Heat sink inner cylinder	Aluminum	13.1	15.8
Base assembly	Edison base insulator	Acrylic, polycarbonate	4.2	5.1
	Inner insulator and adhesive connections	Acrylic, polycarbonate	6.6	8
Driver	PCB, resistors, transistors, capacitors, and diodes		10.1	12.2
Edison Screw	Edison base and leads	Gold-plated steel	12.2	14.8
Total weight			82.7	

Source: Data adopted from Hendrickson et al. [4].

as shown in Table 9.2. Thus, by comparison with the data in Table 9.1, materials footprint could be considerably reduced through appropriate design strategies.

A more efficient strategy to reduce the materials footprint is nanostructuring, which would serve a given function with a lesser quantity of materials. Enhanced material properties via nanostructuring led to slimmer products without compromising on their performance. Lowering the materials footprint will also help reduce carbon footprint and embodied energy. The principal component of LEDs, i.e. LED chip, is typically a few hundred microns; although their thickness could be further reduced, the chip's material fraction is much lower than the packaging materials. Similar is the case with many other semiconductor chips used in electronic devices and solar cells. In most of the cases, the chip's chemical identity remains intact, but the performance is lowered due to increase in contact resistance due to continuous flow of electric current. The loss in efficiency of LED chip could mainly be attributed to the increase in resistance. Desirably, the most efficient way to address this problem is to develop technologies to reverse the electrical resistance of the chip and contacts to their initial ways; unfortunately, there are technological gaps at this moment. Therefore, the most desired way to lower the materials footprint, and consequently, carbon footprint and embodied energy, is to develop much more efficient packaging materials and technologies.

9.3.2 Materials with Multifunctionalities

Products are often made of multiple-materials due to insufficient properties of a single material. Packaging containing polymers and other material's composites

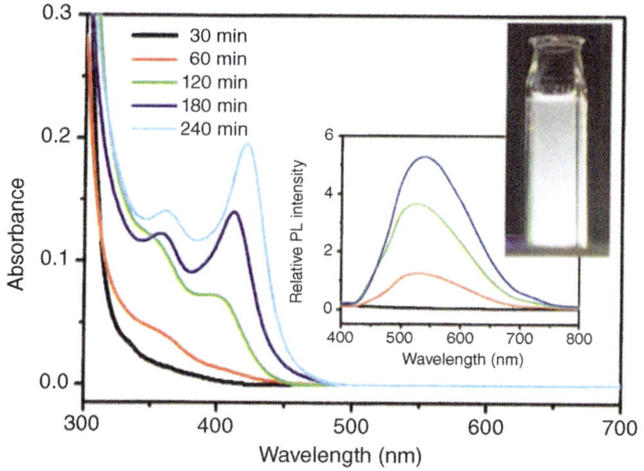

Figure 9.5 Temporal evolution of absorption and photoluminescence (inset) spectra of CdSe QDs synthesized at room temperature. The broad emission spectra are assigned to the white-light emission. Inset shows the white-light emission of the QDs. Source: Reproduced from Jose et al. [6], © 2006, American Institute of Physics.

or compounds (PCB, chip, etc.) is an example of multi-materials in LEDs. Multi-materials system poses tough challenges in recovering materials during end-of-use solid waste management. Therefore, research and development efforts are to conceive and develop simpler materials with multi-functionalities or multi-materials that are easier to separate so that they facilitate ease of identification, sorting, segregation, reusing, remanufacturing, and recycling. For example, by suitably developing a highly conducting material with high optical transparency, the heat sink could be coupled with the optical part (glass or polymer bulbs), which will make the LED lighter and slimmer. Currently, LEDs are fabricated using combinations of different color-emitting diodes and phosphors; developing single materials to enable white-light emission could develop a single material system. Fortunately, white-light emitting II–VI semiconducting quantum dots under ultraviolet light excitation (photoluminescence) have been reported (Figure 9.5). However, no further studies have been reported on developing electroluminescent white light emitters from the corresponding fluorophores.

9.3.3 Materials of Higher Circularity

Currently, the world is only 8% circular, which means that only 8% of the end-used products are prepared for further use through the protocols outlined in Figure 9.2. Only ~20% of over 50 million metric tons of electronic waste generated annually is recycled, although a large reuse market exists. The circularity of LEDs was not effectively addressed during the time of preparing this book, i.e. the circularity of LEDs is currently 0%. One of the primary hindering factors is the unavailability of economically and environmentally viable technologies, which discourages industries from investing in this problem. Performance, cost, properties, and processing are the focus of materials communities and industries thus far in designing and developing materials. Materials circularity, i.e. possibility to be economically repaired, remanufactured, recycled, upcycled, or reimagined, must be an essential materials selection, design, and development criterion for successful circular economy and sustainability. Significant attention is therefore required to develop economically viable and environmentally benign technologies to increase materials circularity. A holy grail is to design and develop materials with perpetual circularity. Materials community must consider circularity in addition to the performance, cost, properties, and processing aspects while selecting, designing, and developing materials.

9.3.4 Materials with Enhanced Durability

LEDs are well known for their durability despite intentional design and selection of materials with shorter lifespans for business profits in recent decades. This is causing growing volumes of solid waste worldwide and associated environmental and human health problems. About 350 million metric tons of plastic are produced annually for short-term use; fossil fuel-derived plastics provide a compelling example, which not only pollute soil and waters but also invade biosystems of living beings, thus posing significant existential threats. Therefore, all materials as well as products should be designed and made for longer life spans, and furthermore facilitate ease of collection, sorting, repair, reuse, remanufacture, and recycling, thus reducing the total solid waste sent into the Earth's ecosystems. Considering the functional diagram and performance data of LEDs, the deterioration of performance of a working device occurs due to decrease in light emission efficiency of the chip, while the packaging parts of the devices are significantly stable. Currently, technologies are not available for selectively replacing the chip in an LED or reenergizing it for continued usage.

9.3.5 Materials with Reduced Carbon Footprint and Embodied Energy

Materials processing is most often energy intensive and a major contributor to greenhouse gas emissions. Consider a typical example of silicone or siloxane used in LED packaging. Using life cycle assessment, carbon footprint estimated during the conversion of silicon into silicone, siloxane, and silane contributes $\sim 10.0 \pm 2.5$ kg CO_2e/kg of silicon. Given that millions of tons of silicon are processed every year for various other applications, this conversion alone contributes to millions of tons of CO_2. Further, because aluminum claims a major share of the LED packaging, reusing an LED by simply replacing the chip will significantly reduce CO_2 emissions, not only from aluminum but also from other packaging materials. Using lifecycle assessment methodologies, it is now possible to estimate the carbon footprint and embodied energy of diverse materials, processes, and products. Yet, majority of the materials researchers have not yet acquainted themselves with these methods and employed them as they advance materials. Immensity of the materials usage and their effects on climate change and human health necessitate the scientific community to develop materials with minimal carbon footprints and embodied energies.

9.3.6 Material Miles

In recent decades, materials and products have been transported over vast distances for their processing, fabrication, and usage; thereby contributing to their increased transportation-related carbon footprint in addition to their extraction and processing carbon footprints. This is typically true for the solid-state lighting industry; most companies are centered in a few technologically developed countries. It is important to reduce material miles by sourcing and making products near where they are consumed. The large-scale production centered in a few locations needs to be reimagined and decentralized, although this will significantly reduce the profit share. Similar strategies could also be adopted for the recyclable/repairable/reusable market. The LED companies must make efforts to improve the functionalities and performances of locally or nearby sourced materials and the device fabrication.

9.3.7 Sustainable Materials from Renewable, Recycled, and Recovered Sources

Nonrenewable sources on planet Earth are depleting, and the Earth is already imbalanced with more human-made materials than all its living biomass. To mitigate this situation, sustainable materials with diverse properties and functionalities need to be conceived and made from renewable, recycled, or recovered sources via scientific advances. For example, biodegradability is an essential characteristic of all renewable materials, and their use in products improves their circularity performance. Organic LEDs (OLEDs) are a promising direction in this case, where the semiconducting properties of organic molecules are fluorophores are employed although solid-state lighting has advantages in terms of longer life, efficiency, and color purity. Wastes are pseudo-renewable; reimagining wastes as resources will not only assure material supply but also provide a better way to address global waste crisis. Therefore, harnessing materials from these sources and making them cost- and performance-competitive is the desired direction to pursue.

9.3.8 Materials with Higher Environmental Benignity

Perpetual recycling or reuse of materials and products is a tall order to realize; a more pragmatic approach is through appropriate science-based green technologies, using which the end-of-use products can be degraded so that they can be absorbed by the Earth's systems in a shorter time with minimal pollution or higher environmental benignity. For example, developing biodegradable plastics or degrading conventional petrochemical-derived plastics in eco-friendly ways. Therefore, the materials community facilitates higher environmental benignity by developing (i) eco-friendly materials from sustainable sources and (ii) Earth-friendly processing and management of end-of-use materials.

9.3.9 Materials with No Adverse Human Health Effects

Metals, metallic compounds and alloys, chemicals, plastics, and composites are the basis of many industries because of the superiority of their properties over materials derived from sustainable|renewable sources. Certain additives in plastics are attributed to adverse effects on human health and well-being. Similarly, metals are cytotoxic beyond a critical concentration. Moving forward, materials research should also be aimed at selecting, redesigning, and producing materials with no adverse human health effects, thus eliminating the social costs of materials.

9.3.10 Materials Enabling Healthy Natural Habitat

The Intergovernmental Science-Policy Platform on Biodiversity and Ecosystem Services noted that up to one million plant and animal species are facing extinction due to human activities such as deforestation, expanding agriculture, industrialization, and urbanization. Materials are a significant part of those human activities. Hence, it is necessary to reimagine materials to nurture a healthy natural habitat. For example, sustainable materials for (i) clean energy generation, storage, and supply, (ii) safe water treatment and supply, (iii) sustainable agriculture, food, and nutrition, (iv) sustainable healthcare, and (v) sustainable homes, transportation, and infrastructure. The materials for solid-state lighting should also be bound to this basic criterion.

The strategies to improve the cost reduction via materials include: (i) improved material utilization, (ii) growth of processing volume of both materials and devices, (iii) improving the collaboration between suppliers for improved material affordability, (iv) improving the materials stability, circularity, and robustness, (v) standardization of material properties via adopting cost-effective and environmentally friendly processing routes.

References

1 US Department of Energy (2009). Solid-state lighting research and development: manufacturing roadmap. https://www1.eere.energy.gov/buildings/publications/pdfs/ssl/ssl-manufacturing-roadmap_09-09.pdf (accessed 12 November 2022).
2 US Department of Energy (2012). Life-cycle assessment of energy and environmental impacts of LED lighting products. Part I: review of the life-cycle energy consumption of incandescent, compact fluorescent, and LED lamps. https://www1.eere.energy.gov/buildings/publications/pdfs/ssl/2012_LED_Lifecycle_Report.pdf (accessed 12 November 2022).
3 US Department of Energy (2012). Life-cycle assessment of energy and environmental impacts of LED lighting products. Part II: LED manufacturing & performance. https://www.pnnl.gov/main/publications/external/technical_reports/PNNL-21443.pdf (accessed 12 November 2022).
4 Hendrickson, C.T., Matthews, D.H., Ashe, M. et al. (2010). Reducing environmental burdens of solid-state lighting through end-of-life design. *Environmental Research Letters* 5: 014016. http://dx.doi.org/10.1088/1748-9326/5/1/014016.
5 Ramakrishna, S. and Jose, R. (2022). Principles of materials circular economy. *Matter* 5: 4093–4512.
6 Jose, R., Zhelev, Z., Bakalova, R. et al. (2006). White-light-emitting quantum dots synthesized at room temperature. *Applied Physics Letters* 89 (1), 013115.

Index

LED Packaging Technologies: Design, Manufacture, and Applications, First Edition.
Luruthudass Annaniah, Mohamed Salleh M. Saheed, and Rajan Jose.